Lecture Notes in Biomathematics

Lecture Notes in Biomathematics

Managing Editor: S. Levin

44

Recognition of Pattern and Form

Proceedings of a Conference Held
at the University of Texas at Austin
March 22-24, 1979

Edited by Duane G. Albrecht

Springer-Verlag
Berlin Heidelberg New York 1982

Editor

Duane G. Albrecht
Department of Psychology, University of Texas at Austin
Austin, Texas 78712, USA

AMS Subject Classifications (1980): 92-06, 92 A 09, 92 A 27

ISBN-13: 978-3-540-11206-8   e-ISBN-13: 978-3-642-93199-4
DOI: 10.1007/978-3-642-93199-4

2141/3140-543210

Contents

PERCEPTUAL MECHANISMS (PSYCHOLOGICAL, BIOLOGICAL, AND ARTIFICIAL): AN INTRODUCTION

Duane G. Albrecht

University of Texas
Austin, Texas

The ability to recognize and categorize objects and events, from the continuously changing patterns of environmental stimulation, is necessarily a fundamental aspect of human and animal behavior: in order to make an appropriate-adaptive response, an organism must first analyze and recognize the particular stimulus encountered. Given the complexity of environmental stimulation, and the essentially infinite number of discriminably different patterns of stimulation, one is naturally left to wonder how such sensory information processing is made possible.

In the past, questions such as this (the classical problems of sensation and perception) were primarily of interest to philosophers and experimental psychologists. During the last few decades, however, the study of "systems" capable of sensory information processing (and ultimately the recognition of complex patterns) has been of concern to at least three different categories of research workers: the experimental psychologist, the neurophysiologist, and the computer scientist.

A large share of the history of Psychology reflects a modern attempt to address the classical epistemological and phenomenological questions of how an object of the world is "known" by a human observer. As Boring, Brunswick, Neisser, and many others have pointed out: the chief intellectual traditions (from British Empiricism and German Sensory Physiology) make perception the first problem, the propaedeutic problem, of experimental psychology. From this perspective, perception is the brain's most fundamental cognitive activity from which all others emerge.

A great deal of research has concentrated on the "psychological mechanisms" of sensation and perception and a variety of different types of models have resulted. One common theme found in most current models is the notion of mechanisms (channels, filters, detectors) selectively sensitive to specific qualities (attributes, features, dimensions) of impinging stimuli. The general idea (similar to Locke's "Mental Chemistry") is that in the process of analyzing a given stimulus, the sensory mechanisms decompose or breakdown a complex pattern into simpler elements (such as, color, size, shape, orientation, etc.). It is then on the basis of this list of criterial features (organized in a hierarchical fashion) that the subsequent processes of discrimination and recognition can complete the task.

Our understanding of the "biological mechanisms" of sensation and perception has been greatly enhanced with the advent of neuroscience techniques such as single cell electrophysiology. In a very real sense, the micro-electrode is capable of isolating the fundamental elements of perception (individual sensory neurons) in a fashion which was never possible using behavioral methodology.

It is now clear that sensory neurons are, in fact, selectively sensitive to certain attributes or features in the environment; or, more precisely, sensory neurons only respond when the qualities of a given stimulus fall within select ranges on a variety of different dimensions (color, size, shape, orientation, etc.). Thus, for example, neurons in the visual pathway of primates are excited by only a restricted range of colors; or auditory, 8th nerve fibers are only excited by a restricted range of tones. One might characterize the goal of sensory physiology as the mapping-out of a detailed wiring-diagram of the sensory mechanisms.

There is a third group concerned with the basic problems of sensation and perception: namely, the group of engineers and computer scientists faced with the task of designing computer-based systems which can "see" and "hear". Much research is being directed towards designing machines which can "automatically" analyze and recognize the myriad of potential stimuli encountered in the environment.

The formidable task faced by the artificial intelligence engineers has, of course, already been completed by "Biological Systems", through the eons of evolutionary trial and error. One might hope, at least in principle, that once the neurophysiologist understands the biology, an engineer can then readily duplicate the mechanism. However, it is clear that the complexity of the biological solution (say the primate visual system) will not be easily understood by the neurophysiologist.

Our current understanding of psychological, biological, and artificial mechanisms, which can analyze and recognize objects and events, is rudimentary at best. The solutions to the general problem (whether discovered in biological systems or artificial systems) are of interest to all three groups. The inter-change of ideas has, of course, been occurring over the decades. In fact, it is often difficult to discern who has influenced whom (the genesis of the notion of "feature-detection", as developed in the late fifties and early sixties, was certainly a mutually reinforcing enterprise).

In an attempt to make a small contribution to this interchange, we held a conference in March of 1979 which brought together leaders in each of these fields to discuss the current state of their research efforts. We asked the participants to bring us up to date on their current projects while keeping in mind the pervasive notion of "feature-detection". The model of feature detection, as a solution to the perceptual problems, is found in every introductory textbook of psychology, neurophysiology, and artificial intelligence. Is the model valid, useful, and parsimonious? The current answers to this simple question are quite complex.

Acknowledgements

The chapters within this volume resulted from a conference held under the auspices of the Center for Cognitive Science of the University of Texas at Austin, with funds provided by a grant from the Alfred P. Sloan Foundation. The conference was suggested and organized by Philip Gough and Stanley Peters. We all thank Dianne Sheftall for her skill and diligence in bringing these manuscripts to press.

THE PAST, PRESENT AND FUTURE OF FEATURE DETECTORS

H.B. Barlow

Physiological Laboratory
Cambridge CB2 3EG
England.

If one knows where ideas have come from and how a subject has developed one gets a much clearer view of which direction you are pointing and what the next problems may be. I shall therefore start by looking at the past history of feature detectors, then I shall talk a bit about my present work, and I shall conclude with a brief guess at the future.

Past history of feature detectors

The idea of feature detectors arose in the early fifties and I think the prime reason was a discontent with the two ideas about the physiology of the sensory pathways that were dominant at that time. The first idea was that of projection maps, which arose from the work of Adrian, Bard, Marshall, Woolsey and others (Adrian, 1947; Marshall, Woolsey and Bard, 1941; see also Fulton, 1949). It was shown that a sensory surface such as the skin, the retina, or the cochlea, was simply mapped topographically onto the cortex. The discovery was important, but physiologists were rather inclined to leave it at that point and say "Well, now we've got the map up into the brain it is for the psychologists to find out what goes on there." However if you thought about it at all, it was unsatisfactory to suppose that the only thing of physiological interest that happened in the retina or in those complicated nerve cells and synapses in the lateral geniculate was simply to map the visual image on the cortex; that would really be rather a disappointing conclusion, for one might well hope to be able to find out more about the kind of operation that goes on high in the nervous system by studying transformations of sensory messages early in sensory pathways.

That was the first idea that seemed inadequate, and the second one was the concept of a receptive field as it had been left by Hartline (1938; 1940a; 1940b). He was the first person to map receptive fields in the retina; this was another important advance, but the aspect of

it which was disconcerting was the enormous size of the receptive
fields and the supposition that they simply responded to the summed
light within their area. Hartline worked on a variety of animals, but
much was done on the frog, which one knows is a very visual animal
capable of a high grade of visual performance, and yet the fields
which he mapped were absolutely enormous, covering ten or twenty
degrees in visual space. When I started working on the frog's retina I
remember explaining, actually to Alan Hodgkin, what I was trying to
do; I said that I didn't see how frogs could possibly read and write
if they had summating receptive fields that size, and his answer was,
naturally enough, that they don't read and write. But they do catch
insects and they depend on their eyes for their living, so I felt
their eyes were bound to have good resolution.

Those were the reasons for being worried about the physiologists'
description of what went on in the visual pathway, but there was a
positive stimulus for thinking in a different direction, and this came
from the ethologists. About that time Tinbergen (1951) and Lorenz
(1952) were putting forth the idea of specific releasers for
particular pieces of behaviour in animals. We were hearing of examples
like the red dot on the herring gull's beak toward which the young
make their pecking movements, and the various releasers of the
different items of behaviour in the stickleback. Now physiologists of
the central nervous system were not at that time thinking atomically;
they were frightened at the prospect of trying to explain a complex
piece of behaviour of the whole animal in terms of what single cells
do, and nobody would have dreamed of trying to do it. But the idea of
accounting for little items of behaviour in terms of single cells
specifically tuned to red dots and things like that was quite a
different matter that did seem to be a possibility.

This possibility should have been evident when I started looking at
the frog's retina but I actually had a simpler idea in mind. If you
measure the psychophysical threshold of a test stimulus as a function
of its size, when it's very small it follows Ricco's law: that is, the
threshold is inversely proportional to the area and it is as if it was
summating all the light within a particular region. But when you get
to larger areas it obeys Piper's law, which says that the threshold is
proportional to the square root of the area; alternatively you can say
that threshold is inversely proportional to length of the edge. So was
it possible that the units which had these enormous receptive fields
were not simply summating light all over that large area, but were

abstracting the edge and responding in proportion to that? Were they "edge-detectors", not just transducers of total light energy?

To test this I did the experiment on a frog retinal ganglion cell shown in Fig. 1 (Barlow, 1953). The log of the radius of the spot is plotted horizontally, and log(1/threshold) vertically.

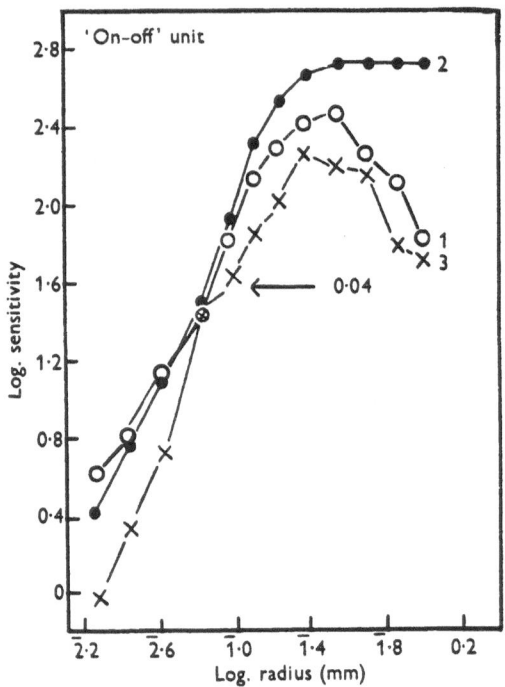

Figure 1: Test for "edge-detectors" by measuring the threshold of an "on-off" ganglion cell of the frog's retina to spots of varied radius. Curve 2 was calculated from a map of the whole receptive field on the basis of summation of sensitivities of the parts covered by the spot. Curves 1 and 3 were experimental measurements of threshold done before and after the mapping. If the unit had done simple summating the points would have followed curve 2; if it had done "edge-detecting" they would have risen with unit slope on these coordinates. They do not fit either hypothesis well, but their decline at large radius revealed the presence of the inhibitory surround (from Barlow, 1953).

I was hoping that the points would follow a line of slope 1, indicating that threshold was inversely proportional to the length of the edge. The line numbered 2 was the alternative prediction, that sensitivity was simply the sum of the sensitivities of the retinal regions covered by the spot. Well, however hard I looked I could not make that into a line of slope 1, so I was a bit distressed, but I got some comfort from the decline in experimental points at large radii, because if they were edge-detectors one would expect sensitivity to decline when the edge of the spot fell outside the receptive field. And then it occurred to me that there was a very much simpler explanation of these things; it might be simply that there was an inhibitory effect from the surround, and indeed there was, as Fig. 2 shows. If you put the spot in the centre you get a good response at both "on" and "off." But if you turn the other spot on and off at the same time you get very much less. The original idea about edge detection by means of some mysterious process that abstracted the edge turns out to be wrong, but on the other hand, something very similar is achieved, though on a coarser scale of distance, by having the inhibitory surround.

For me, this evidence for lateral inhibition was the beginning of the idea of feature detection, and it developed further in the direction of the lead given by Lorenz and Tinbergen. If you really want to stimulate these cells vigorously you will find yourself delivering a type of stimulus to the receptive field which is similar to the stimulus that activates the frog into violent feeding behaviour in real life. I remember on one occasion qualitatively testing various spot sizes and rates of movement and flicker to see what drove a cell best, then doing sums for getting the same angular subtense of a stimulus in the real world of a frog, and finally taking a piece of paper mounted on a wire down to the frog-tank in the basement. The effect was dramatic: the tank was transformed from a torpid mass of lifeless bodies into a cage of active, prey-seeking beasts. So the idea of explaining the innate releasers of Lorenz and Tinbergen in terms of the functions of single nerve cells did not seem impossible, and the surprising thing was that a substantial part of the lock-and-key mechanism appeared likely to reside in the retina. Of course one now knows from the work of Ingle (1968) and others that it was too naive to suppose that a single class of retinal ganglion cell was the only thing that determined feeding behaviour, but nonetheless the retina contains selective filters for fly-like objects.

This work received publicity from a surprising source, the U.S. Navy; William Neff was ONR representative in London and circulated an account of a lecture I gave in 1952 on "The Psychology of the Frog's Retina" to US Vison labs. And there were others talking about similar things, such as Waterman, Wiersma and Bush (1963, 1964) with their

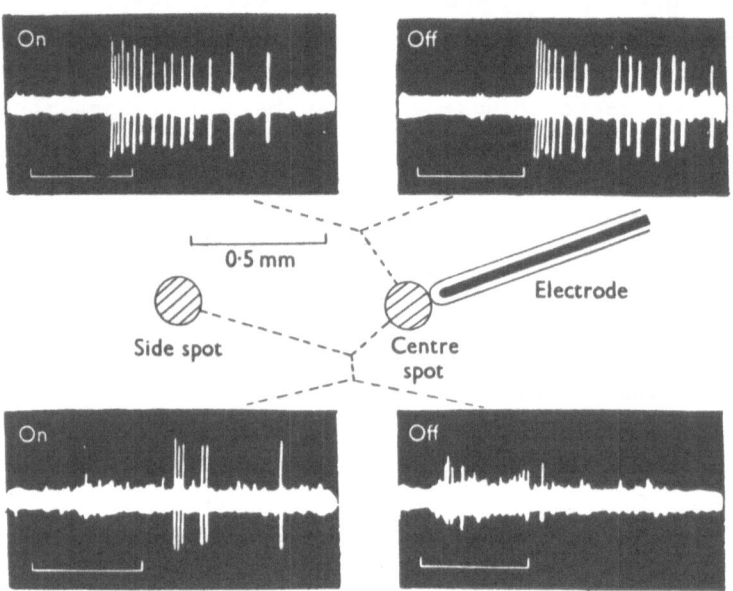

Figure 2: Lateral inhibition in the frog's retina. One spot of light was placed near the centre of the receptive field of a ganglion cell; this gave vigorous bursts at "on" and "off", as shown in the records at top. A second spot was placed almost 1 mm away, and this inhibited the response from the first spot when turned on and off synchronously with it, as shown in the lower records. This shows the presence of a silent inhibitory or suppressive surround; unlike in the cat, the second spot gives no response at either on or off by itself (not illustrated). Time calibration is 200 msec. (From Barlow, 1953).

work on crustaceans. But I don't think Feature Detectors really caught on until 1959, when Lettvin, Maturana, McCulloch and Pitts (1959) published their famous paper on the frog retina, and when Hubel and Wiesel (1959, 1962) started publishing their work on the cat cortex.

It is an interesting question how much any of us were actually looking for what we found. One person whose name I haven't mentioned yet, who certainly was selling the idea of feature detectors very prominently around that time was Oliver Selfridge (1959) with his "Pandemonium" model. This is really a set of feature detectors arranged in a particular way, and of course the "Perceptron" of Rosenblatt (1959) was also a serial hierarchy of feature detectors.

Hubel and Wiesel always denied strenuously that they were looking for anything; they just found out what was actually there, and this was a very good line to take, especially as the rest of us could not help looking for what we hoped to be there. I am not certain this story of Hubel and Wiesel's discovery is true; it is said they were intent on mapping out the receptive fields in the cortex using little spots or light, or black dots, and for this they borrowed from Steve Kuffler a set of black dots on glass slides that we had used down in Johns Hopkins. But in the journey from Johns Hopkins up to Boston they had been cracked, so that the slides had lines on them as well as the dots, and they observed that a line of the right orientation was a much better stimulus for their cortical neurones than any black dot. If true, that should be empirical enough for anybody, and I think it should bring home the fact that you need the concept of feature detectors to describe the way neurones in sensory pathways behave. Orientational selectivity cannot be an uninteresting artefact.

Before leaving the past let me illustrate two other types of feature detector with which I have been involved. Fig. 3 shows a directionally selective cell of the type Levick, Hill and I found in the rabbit's retina (Barlow, Hill and Levick, 1964); Maturana and Frenk (1963) described similar units in the pigeon retina. First, if you map out the receptive field with a stationary spot there is nothing to tell you of its most interesting property; everywhere inside the O's just gives a brief burst at both "on" and "off." But when you test with a moving stimulus you get really dramatically asymmetric responses. For the spot moving in one direction you get hardly any, just two, impulses. But when you move it in the opposite direction you get a vigorous, sustained, burst containing, in this case, 79 impulses. In other directions you get intermediate numbers. Incidentally, we came across this really striking property of cells entirely by accident; we were not expecting it or looking for it at all.

The other type of selectivity I want to illustrate is for disparity in cells of the cat cortex. Figure 4 (from Barlow, Blakemore and Pettigrew, 1967) shows two units, one on the left and one on the right. You have to imagine the cat looking at the plotting screen with its eyes slightly diverged, so that for the left unit, to stimulate the left eye you have to put the object as shown at the top and to stimulate the right eye it has to be as shown in the second row. So you measure the separation of those two stimuli and move them together

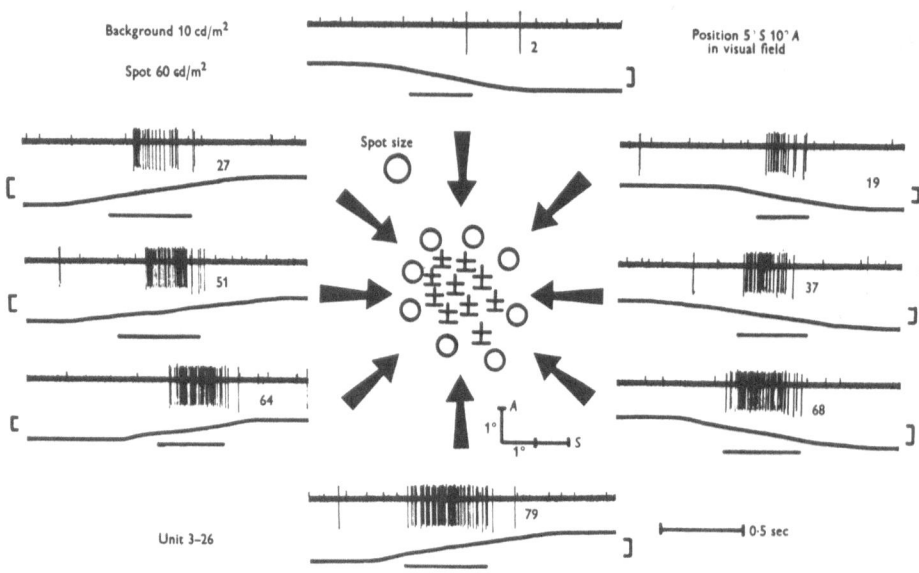

Figure 3: Directional selectivity in a retinal ganglion cell of a rabbit. The receptive field was mapped with a stationary spot and gave responses at "on" and "off" at all points within the O's, and no responses outside. When tested with a moving spot it gave a vigorous bust of 79 impulses for anterior movement (upward in the figure), but only 2 for posterior movement. Other directions gave intermediate values. The slanting line below the trace of impulses monitors the movement of the spot, the calibration marks indicating 5 degrees in the visual field. Other units respond selectively to posterior, upward, or downward movement. (From Barlow, Hill and Levick, 1964).

and then you get a big response, 17 impulses on this trial, much bigger than the sum of the two eyes separately. If you do it at the wrong separation you get fewer, less even than with either eye alone. The point to notice is that for the optimum response, the stimuli have to be more than 6 degrees apart on the plotting board; that is not a true disparity of 6 degrees, but the sum of the disparity and divergence. The unit to the right is rather similar, but for optimum response the stimuli need to be just over 3 degrees apart. So these two units will respond to objects at different distances and are examples of disparity selectivity.

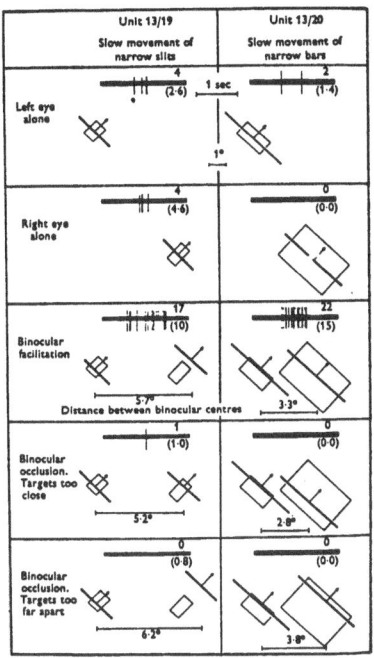

<u>Figure 4:</u> Disparity selectivity in cortex of cat. Two units
are shown, one of which showed binocular facilitation when
the moving targets for left and right eyes were separated by
a distance corresponding to 5.7 degrees on the plotting
screen, the other when separated by 3.3 degrees. The eyes
are slightly diverged in the paralyzed preparation, but if
converged by an amount appropriate for the first unit to
respond optimally to a real object, then an object capable
of stimulating the second unit would have required 2.4
degrees additional convergent disparity, and would therefore
have had to be much closer. The number above each record
shows the number of impulses in that trace and the number
below in parentheses gives the average in 5 repetitions.
Notice that there is no overlap in the range of disparities
for which the two units respond. (From Barlow, Blakemore and
Pettigrew, 1967).

In this case Pettigrew, Blakemore and I deliberately searched for
disparity selectivity; we thought it would be there and it was. But
even though theoretical reasons for expecting feature selectivity have
been advanced, the overwhelming fact is that you need the concept of
feature-selective detectors just to describe the findings when
recording from neurones in sensory pathways. Whatever their future,
our knowledge of feature-detectors represents a substantial addition
to our previous knowledge of topographical mapping and of non-pattern-
selective receptive fields.

Feature detectors in human psychophysics

Most people would accept the proposition that, if frogs, cats, rabbits and monkeys have feature detectors in their visual systems, so do humans. But how do we obtain evidence for their operation in human sensation and perception? There are a host of problems in relating neurophysiology and sensation. For instance it can be said that neurophysiology gives us details of the hardware, whereas psychophysics only tells us about overall performance, which can be achieved by many different types of hardware; or that neurophysiology describes local operations performed on the image whereas the most important aspects of sensation are global; some people might even go so far as to say that the psychological approach can tell us nothing about mechanism, though I think that is falsified by the successes of psychophysics in elucidating the mechanisms of colour vision, and the limits of sensitivity and resolution. I do not claim to have overcome these very big problems, but it is my belief that they can be overcome if one approaches the problem the right way, and I think the right way has been shown to us by the successes I have just mentioned. So my objective in the work that I shall now describe was to make measurements of perceptual performance of a type that must tell one something, at least, about the necessary hardware, and I have picked a problem that certainly involves global, not purely local, operations. I hope it will become clear where feature detectors come in later.

There are two key ideas in this approach. The first is to use a measure of performance that will allow comparisons to be made at quite different tasks. There is no problem if the task is always very similar. Suppose, for instance, that one believes the visual system has "bar detectors" for vertical and horizontal bars, but not for obliques, which is probably the case in rabbit retina (Levick, 1967); if this was the case in humans, one should be able to get evidence for it by measuring a subject's contrast sensitivity for bars at different orientations; it should be poor for obliques. But it would be different if one was varying the length or width of the bars, or comparing bar-detecting with movement or disparity detection, because changing the task in these ways would make it physically more difficult or easier. How then is one to compare performance at totally different tasks? The only answer I know is to measure each performance by comparison with an ideal instrument for performing that task. By making use of concepts derived from statistical decision theory and information engineering this can be done for a good many perceptual tasks, and it yields a figure for efficiency; it answers the question "What fraction of the information available for performing this task did the subject actually use?" Naturally we expect, using this efficiency measure, to find that subjects are better at some tasks than others.

The second key idea of this approach is to search for tasks that human subjects can be shown to perform efficiently, using these absolute measures. Now loss of efficiency is irreversible; if information is not collected, or is thrown away, no kind of central processing can make good this loss. So if a subject performs efficiently his physiological mechanisms must be preserving the requisite information, and the hope is that one can find tasks that optimally match his physiological mechanisms, thereby defining their most important functional property.

Notice that the efficiency measure can, in principle be used on a neurophysiological preparation; in the future years we may be able to say "Human subjects detect bars with X% efficiency; we find single neurones in monkey visual cortex that perform with equal efficiency, and we therefore believe that human subjects detect bars by using such feature detectors in their own brain". Strictly, we should say "we therefore see no reason for disbelieving", but the message is the same.

An important characteristic of this approach is the assertion that there is a limit to how well a perceptual task can be performed. A lot of psychophysical measures, such as those involved in acuity and colour matching, relate to the physical aspects of the stimulus; you try and explain how well you do them in terms of the quality of the retinal image or in terms of the photo-sensitivity of the pigments and so on. In the type of problem I am going to deal with here we are not concerned with physical limits of that sort, so the first thing to decide is what else might limit the performance of feature detectors, and the overall performance of the whole animal, when making a higher level perceptual decision.

I shall argue that in many cases, perceptions are limited by noise and not by insensitivity of the central apparatus. I think most people are familiar with the fact that if you have an instrumental system designed to measure some particular quantity, it may fail to detect an input for either of two reasons. It may be too insensitive, because it does not have enough amplification, or because the pointer of the output galvanometer has too strong a return spring so that it is jammed up against its stop, or something like that. On the other hand, the limit may be of a different kind if, even in the absence of the stimulus that it's designed to detect, the output pointer or indicator is showing signs of continuous activity. In that case the performance of the system will be limited by signal-to-noise considerations. And I'm going to advance some arguments that that is in general the case with perception.

Most people start with a prejudice in the other direction because of one's subjective impressions. When one looks at an array of photographs, for example, and picks out the face of somebody one knows, one doesn't have the impression of a whole lot of false recognitions going on in one's mind, then the correct one being selected, with some doubt, from a number of false ones. The process seems more like the pointer on a meter coming up to a definitive reading. And likewise with almost all sensations, when you look at them superficially they seem to have this character of being rather certain judgements without an element of background uncertainty. However, that's not the case when you start looking in more detail. For example at the absolute threshold, if you start encouraging people to respond without being quite so certain, you can drive their criterion so far down that they will make almost any desired proportion of incorrect identifications (Sakitt, 1972). The initial impression that perceptions are not perturbed by noise does not stand up to closer examination. That's the first argument.

The second one is rather general and difficult to make convincing. It is simply that one's got to perform some complicated operation on an input which has spontaneous activity; the sensory nerve fibers are not quiet, they're buzzing away like mad most of the time. And it is very difficult to imagine <u>any</u> operation that could be performed on that input that wouldn't occasionally give a false response. There must be false identifications of features, because one could not design a mechanism which would be entirely free from errors.

And the third point is a demonstration. Figure 5a shows the same set of words, but each has been carbon-copied an increasing number of

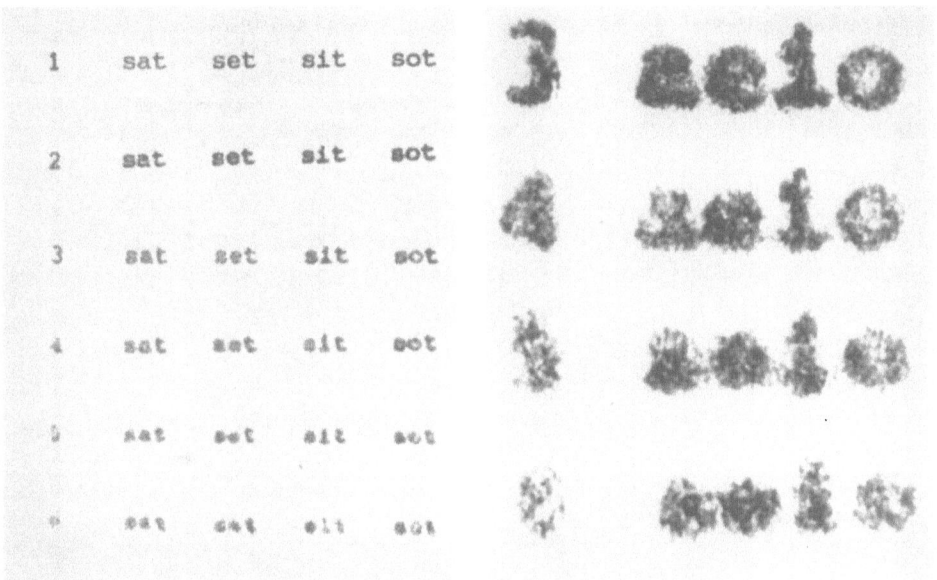

Figure 5: To the left (A) are shown the first to the sixth carbon copies of the same set of words; after the third they become hard to identify. It might be thought that enlargement would improve matters, but the enlargements of the central vowels at the right (B) show that this is not the case. It is the noise introduced in the copying process that impairs perception, and in such cases enlargement, amplification, or contrast enhancement do not improve performance.

times. The fifth or sixth copy becomes virtually impossible to identify. If they were small, you might think that it was merely an amplification problem;if the size could be increased a bit it would be possible to identify them. But that is not so. At the right (figure 5b) are the same letters enlarged and as you see it is obviously not a problem of magnificaiton, but of the background noise in the system. It is plausible to suppose that many other recognition and identification problems are also limited by noise and not by insensitivity.

Now in order to be quantitative about this one must look at the question how well discriminations of signal from noise are performed; it is a statistical task. I wanted to look into a type of perception which was more complicated and must involve higher level processes than ones involving resolution, sensitivity, or colour. The task I chose is shown in Fig. 6.

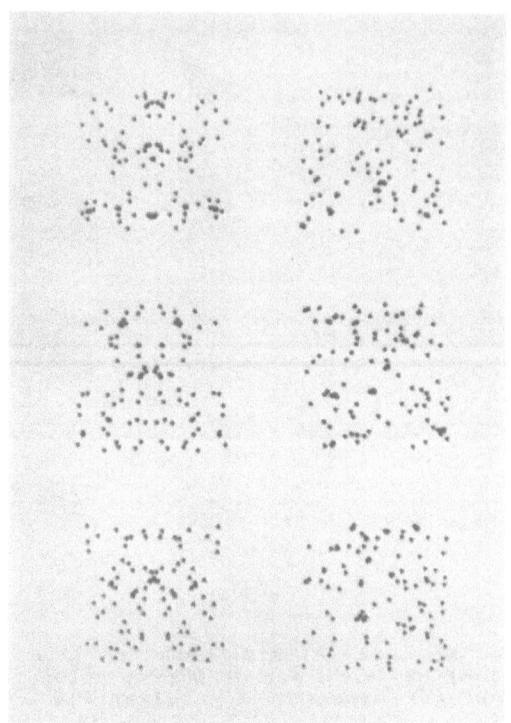

Figure 6: Pairs of computer-generated dot patterns that were either symmetric (left) or without symmetry (right). The subject's task was to assign an example to one population or the other. In these examples he would make no errors, but the limits of the symmetry-detecting mechanism can be explored by making the task more difficult.

These are photographs of the graphics screen of a computer. The three on the left are all symmetric about the mid-line. The three on the right are not. One can feel absolutely certain there is not any simple optical or photochemical explanation for how this is done; you are quite out of the realm of things which are limited by the physics of the eye.

It is clearly more complicated than many simple detection processes in another way because it demands an associative process. If half of a picture is covered you couldn't possibly tell if it was symmetric or non-symmetric. So there must be some way of detecting an association between corresponding positions.

And thirdly, I recalled the days when I tried to persuade psychologists in Cambridge that it was meaningful to talk about the psychology of the frog's retina. They used to come back and say, "Well, that's all very interesting but it's not real psychology at that level; that's ethology or mere physiology; you've got to show how the interesting problems like those of Gestalt Perception are performed before you're really saying anything of the least importance to us". I thought symmetry would be an interesting problem to look at for this reason too. We may not find the mechanism by doing psychophysical tests, but we can at least define how well it works and under what conditions.

The method was as follows. The subject sat in front of a computer graphics screen and for a familiarization period, which could actually be as long as he liked, he simply pressed "zero" or "one" to get either a sample of one of the right hand patterns or one of the left hand patterns of Fig. 6. As you can see, if you did that with those patterns, it would not take you long to realize that when you press the one you always get symmetric pattern. The subject is given every opportunity to learn and verbal instruction was included if necessary. Then the computer gives him a hundred unknowns and it is the subject's job to classify them. If it had been a task like that of Fig. 6, he would have had no problems; he would have got 100% right, but there are ways of making it more difficult and finding out the conditions where the symmetry detecting mechanism breaks down.

These experiments were done with Barnie Reeves, who is now in Oxford, and have recently been published (Barlow and Reeves, 1979). The first thing we did was to find out more about the elementary properties of symmetry detection. From the literature we were not clear whether the axis of symmetry had to be central and vertical. Well, the answer is that it doesn't matter, though one performs best with the axis central and vertical. I should mention that in these experiments the picture was only flashed on for about 200 msec, so the person could not tilt his head or anything like that, and in most cases could only use one fixation pause. It is of course important that you can do this kind of task in a brief flash with no particular difficulty.

We also did experiments in which there was a fixation point on the screen, and the picture appeared either centred, or to the left, or to the right. The subject had no way of anticipating which place it would

occur and could not move his fixation in anticipation. The result was
that performance on the centred ones was higher than on the displaced
ones. So it is true that you do it better with the axis on the
vertical mid-line, but you can still do it when the patterns are
displaced up to three degrees into the periphery. These patterns were
two degrees across, so no part was falling in the centre of the field
of vision, and yet the performance, though it was decreased compared
to the centred one, was still quite good.

Obviously the physiological mechanism required to detect symmetry
about any orientation and any position in the visual field is vastly
more complicated than that required if it had turned out that one
could only do it with the axis vertical and in the mid-line. For the
latter operation one stereotyped mechanism might be enough, but we now
know the mechanism can perform quite a versatile task.

To achieve the aims I set out above we must now try to express in
absolute terms how well this task can be done. To do this we must
compare the human performance with the best possible performance, and
it is only then that we may get evidence that the task is being done
so well that we might tentatively conclude that there is a specific
mechanism, a "feature detector" for detecting symmetry. How are we
going to decide what limits the detection of symmetry and how closely
our subjects approach this limit?

These dot patterns were made with a computer, so we know exactly
how they were made and it is feasible to say how you should set about
testing whether a pattern is from the symmetric set or not. It would
actually be dead simple: you take any dot, calculate, the coordinates
for the mirror dot and look through the list to see if any of them
coincides with that. And if you found one mirror dot you could say,
"Well that couldn't arise by chance, therefore the pattern must be
from a symmetric set". It could hardly arise there by chance because
these dots are placed with high accuracy, 10 bits for each dimension
approximately, so there are one million possible positions for a dot,
and the chance of one of the 100 dots occuring at a prespecified
position is very small, unless it has been deliberately put there. As
soon as one gets evidence of even a single symmetric pair one can say
with reasonable certainty that the pattern is from a symmetric set.
However the situation would be very different if the dots were placed
with lower accuracy, for then pairs could arise by chance in
asymmetric patterns. Notice, moreover, that if the visual system is

not capable of using all the accuracy in the display, then our equipment would be better than the subject needs, and the above argument would also have given a false impresson of how easy the task is. Therefore what we did was to decrease the accuracy

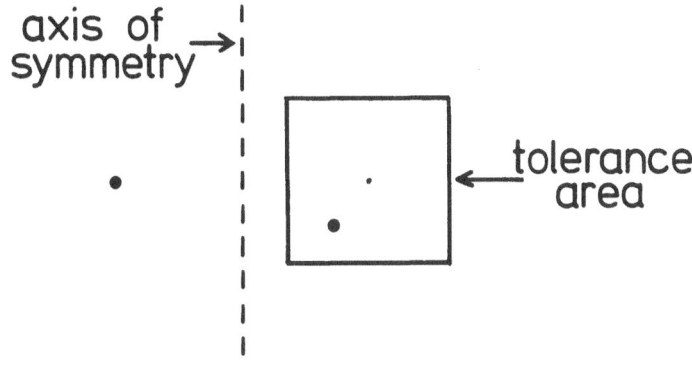

<u>Figure 7</u>: Generating smeared or imperfect symmetry. The position of the dot at left was first selected by the computer; then instead of placing the paired dot at the exactly symmetric position, it was placed at a randomly selected position in a tolerance area. The size of the tolerance area could be varied in order to find how performance at detecting symmetry was affected. (From Barlow and Reeves, 1979).

with which the dots were mirrored. Instead of placing a dot at the exactly symmetric position we placed it in a tolerance area as shown in Fig. 7. Then we varied the tolerance area, the aim being to match the properties of the generating system to that of the detecting system.

Figure 8 gives an impression of what this "smeared" symmetry looks like. It looks as though the symmetric pattern had been dropped on the floor and slightly damaged. The usual task was not to distinguish smeared symmetry from perfect symmetry, but from a completely random

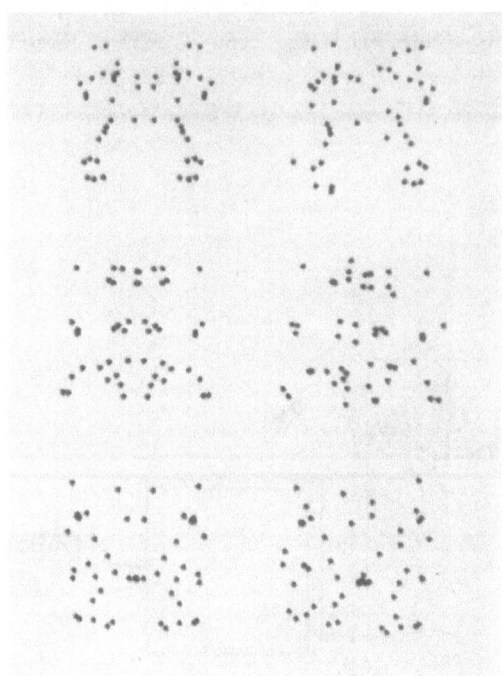

<u>Figure 8</u>: Three examples of exactly symmetric patterns (left), and the same patterns in which the symmetry has been "smeared" (right) by the method shown in Figure 7.

pattern. Figure 9 shows how d' varied at different tolerance ranges. The surprising thing is that you can tolerate a large amount of smearing, of inaccurate pairing, and still perform quite well. The tolerance box in Fig. 7 can be 10 minutes in each direction, and it has hardly any effect on the performance. You should compare that figure of 10 minutes with two-point resolution of one minute, or the accuracy of vernier alignment, which is also a judgement of position, of the order of 10 seconds of arc. The inaccuracy which you can tolerate and still detect symmetry is very large indeed. Symmetry detection does not need a high resolution system.

Now you remember I said just before that if the dots were inaccurately placed one would have problems in the ideal method. Suppose for example that we use 10% accuracy of placing the dot, then you have got only 100 meaningfully different positions to put a dot, and then when you put down 100 at random, some of them will undoubtedly appear in a range which qualifies as being that of a pair.

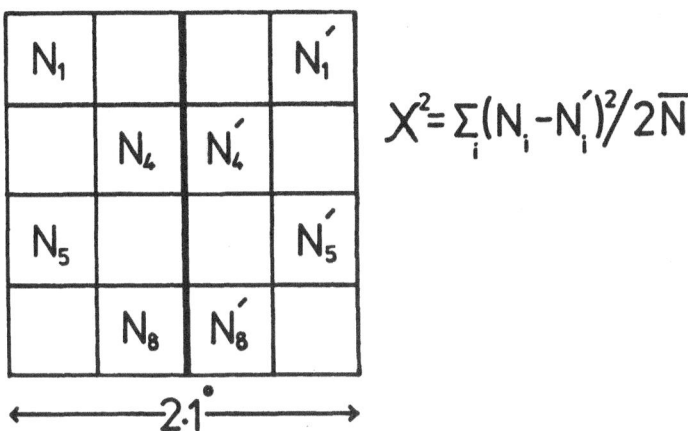

$$X^2 = \sum_i (N_i - N_i')^2 / 2\overline{N}$$

Figure 9: The subject makes errors in detecting "smeared" symmetry, and it is possible to calculate d' for discriminating between the two populations of computer-generated patterns. The experimental values, d$_E$ are shown here for two subjects and a range of different tolerance ranges. The continuous line shows results of basing discriminations on counts of the number of pairs that would qualify as being placed according to the procedure of Fig. 7, since this is the best possible performance; the scale for this line is at right, and is double the left scale. For tolerance ranges above 10 minutes the curve fits the points; hence d$_E$/d$_I$ = 0.5, and statistical efficiency is 25% (see text, and Barlow and Reeves, 1979).

So how can you tell which patterns come from the paired set and which come from the completely random set? The best way of doing it would be to go through all the pairs--4.950 for 100 dots-and count those which could have been placed as pairs, thus obtaining the total number of such qualifying pairs. In the symmetric sets some pairs have been placed deliberately, so the total number of qualifying pairs is going to be increased. But of course the number is going to be very variable, so you have got a background of noise of "false pairs" in determining symmetry.

When the tolerance range is small the calculation of the expected number of qualifying pairs is straightforward, but when the tolerance range is large it becomes tricky because dots are then spreading outside the original area and the average density is non-homogeneous. So we did a computer simulation in which a series of patterns was generated, and then the computer sorted through the pairs and decided which qualified as symmetric. The line in Fig. 9, using the scale on the right, shows what value of d' the computer could achieve using this count of qualifying pairs as the decision variable. It shows no plateau with small tolerances as the experimental points do, because of course the computer can go on using whatever accuracy is available. But for large tolerances the theoretical and experimental points can be made to agree by plotting the theoretical points on a different scale from the experimental. In fact the scales differ by a factor of 2, and that means that the ideal d' is twice the experimental d'. The efficiency of utilizing the information present in the patterns is obtained by squaring the d' ratio, so in this case the figure is 25%.

For tasks of this type I have shown you that human subjects use about 25% of the statistical information. This is a fairly high figure, and we started wondering what kind of mechanisms might achieve it. The first attempt at a model is shown in Fig. 10.

We supposed that the mechanism does not work on the exact position of every dot, but counts up the number of dots in big squares and compares each with the number in a symmetric square. If the pattern was symmetric these numbers should be equal. The mechanism then calculates a statistic, which is a chi-squared test for the absence of symmetry. If chi-squared for a particular pattern is above a certain figure the mechanism says it is not symmetric, if it is below it says it is. With a computer simulation of this test we get the result shown in figure 11. The continuous line is the ideal method as described above, but now plotted on the same scale as the other points so it is much higher. The points are the experimental quantities as before, and the dotted curve is the prediction of the chi-squared model. It seems to fit rather well. And it is more convincing lying well below the ideal curve, one could argue that a lot of different methods would all have given the same result.

We should maybe have stopped at that point, but we thought we should do some additional tests of the model. The first one was to constrain the generation of the dots so there were always an equal number in each of those big squares. Well, if the model is correct,

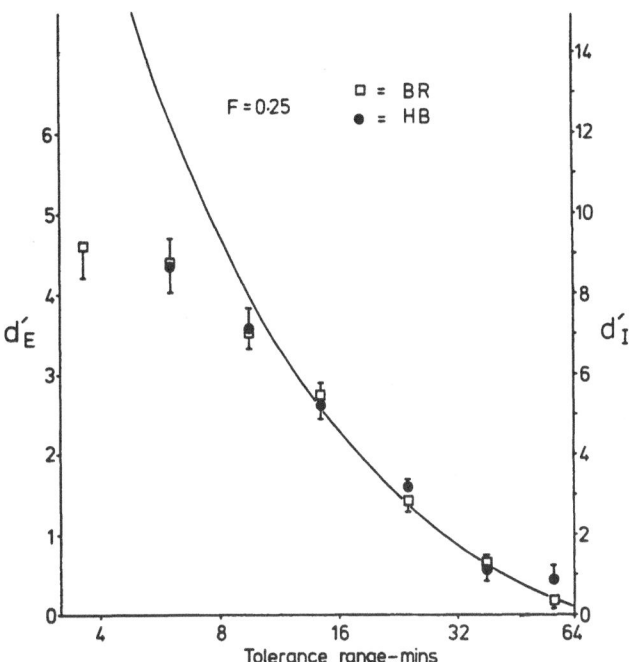

Figure 10: This first model for the symmetry detecting mechanism uses the number of dots in 16 large subdivisions of the pattern, rather than the positions of each dot. The counts in symmetric subdivisions are compared, and x^2 calculated for the hypothesis that there is no symmetry. The performance of this model is shown in Fig. 11 (from Barlow and Reeves, 1979).

that should have abolished all appearances of symmetry, because the model says you only use the number of dots in each of these squares. That's perhaps a little bit harsh because of of course one could have had more squares overlapping without destroying the essence of the model, but the results were devastating because one could perform this task almost as well as the original one. So that wasn't very encouraging. We did some more tests; for example, we took a generated pattern and then re-randomized the dots within those squares. That again should have had no effect on performance according to the model, but it did. It reduced the efficiency from about 25% to about 1%. So again the model looks bad. We then tried replacing the array of dots in a square by a single dot whose area or whose intensity varied according to the number in that square. We did rather better on that; you could reach about 10% efficiency, but this is not as high as the 25% for the original pattern.

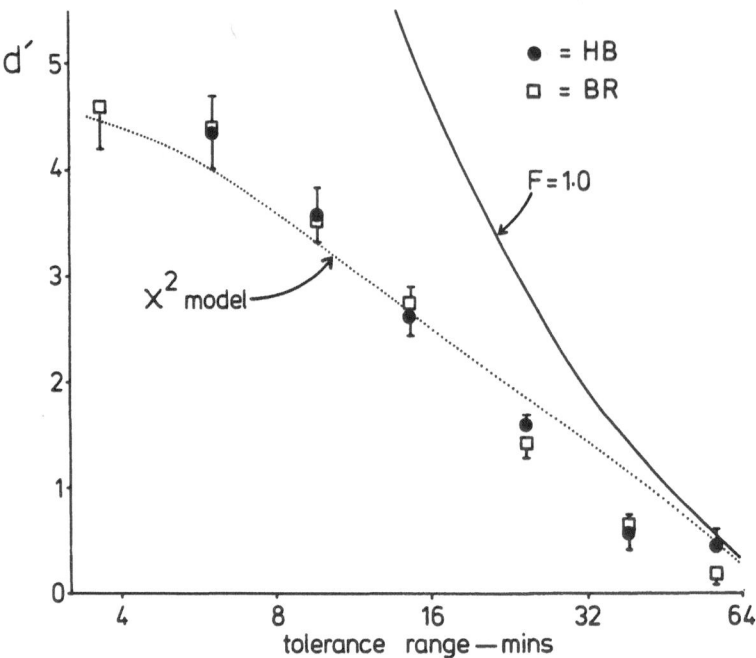

Figure 11: The dotted line shows the performance of the x^2 model of Fig. 10. The experimental points are reproduced from Fig. 9, and the continuous line is the ideal performance curve from Fig. 9 plotted here on the same scale as d_E'. The model describes the performance but subsequent tests show that higher efficiencies than it predicts can be obtained; it therefore requires modification. (From Barlow and Reeves, 1979).

We were discouraged with the model at that point in spite of the good fit of Fig. 13, and decided to test out the essential features one at a time. Of course the most important feature of this model is the fact that it works at low resolution. Instead of working from the exact position of each dot, it uses the total number within a certain area. We thought it would be a good idea to try the effect of blurring these symmetric patterns; if the mechanism works at low resolution, it should do little to mar performance. So we placed a diffusing screen in front of the oscilloscope and Fig. 12 gives an idea of what the patterns then look like. These are the same patterns as those in Fig. 8, but seen through the diffusing screen. They are rather beautiful and one is strangely fascinated looking at them. They are somewhat reminiscent of cross-sections of the brain at various levels, but I hasten to say that I don't think that's significant.

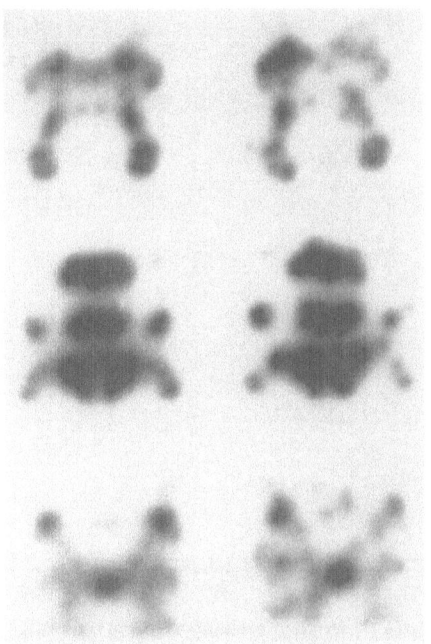

Figure 12: The appearance of the patterns of Fig. 8 when viewed through a diffusing screen. Subjects are better able to discriminate symmetric patterns under such conditions (see figure 13).

Figure 13 shows measurements of performance as a function of the width of the line-spread function produced by the screen. As expected we found performance was degraded if you have severe blurring, but the peak values are obtained with a full width at half height of 20 to 30 minutes, and even a blur of width 1 degree only reduces performance from a peak d' value 3.5 to a figure of 2.5. The fact that you require severe blurring to degrade performance seemed to be in favour of the model. But as you can see, these peak values are actually higher than with no blur at all. That was certainly not predicted by the model; any degrading effect such as that produced by the diffusing screen should mar performance, but clearly it does not.

Realizing that detection gets better with mild blurring, we repeated the previous experiment, in which we had measured performance as a function of tolerance range, and this is shown in Fig. 14. The results for small smears are unreliable, because too few errors were made, but the ones for larger smears are reliable and now the line for

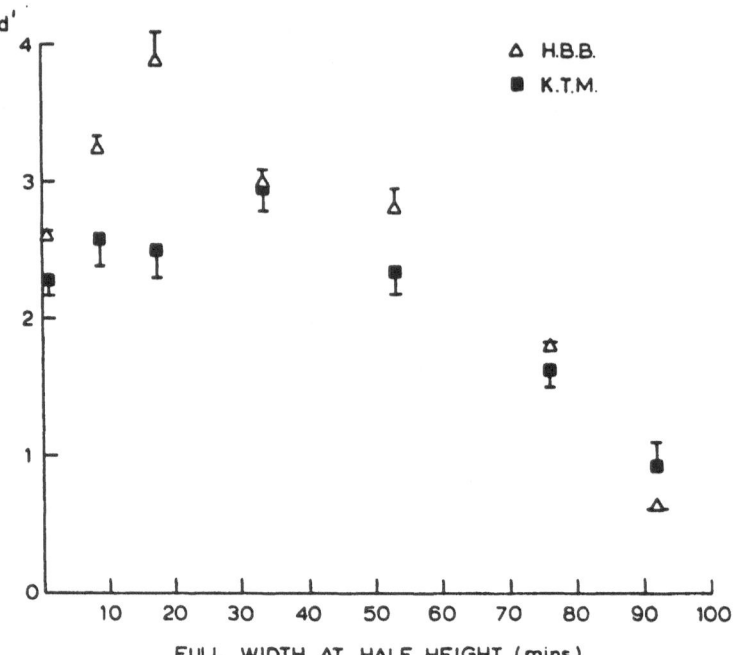

Figure 13: The effect of the diffusing screen on discriminability of symmetry. Patterns with a tolerance range of 15 mins vertically and horizontally had to be discriminated from random patterns, which can normally be done with d_E 2.5 approx. When blurring was severe, discriminability was impaired, but this only occurred when the line-spread function produced by the screen had a full width at half height of about 50 minutes or more. For moderate blurring performance was not degraded, as expected from the hypothesis that symmetry detection uses a low resolution. What was unexpected was the improvement in performance by moderate blurring, up to 30-40 min full width at half height.

50% efficiency goes through them; thus the diffusing screen has increased efficiency from 25% to 50%. That rules out completely the previous chi-squared model, which matched the points but was only 25% efficient. The virtue of using the efficiency measure is that if you get a measured performance which is above that of the model you know the model must certainly be wrong.

Well that's where we are on that particular problem. The conclusions to be drawn are that symmetry detection depends on low resolution information, but the chi-squared model of Fig. 10 is not correct and needs modifying. We haven't done that yet, but we do now

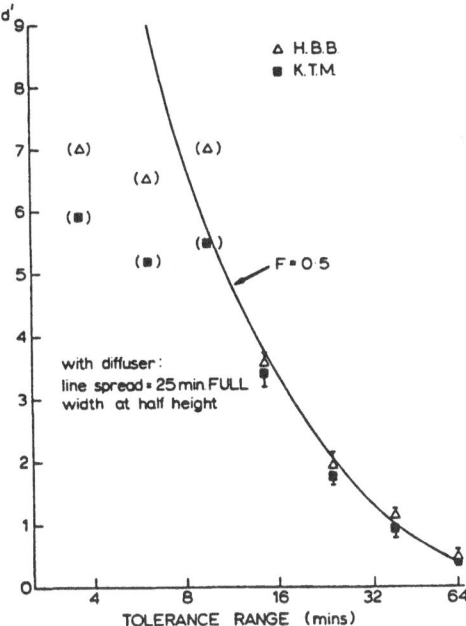

Figure 14: Discriminability of random from symmetric patterns with varying smears when viewed through a diffuser yielding a line spread function of 25 min full width at half height. Points in parentheses are unreliable because too few errors were made. For tolerance ranges of 15 mins or more the points have an estimated standard error as shown, and they indicate an efficiency of 50% under these conditions.

have quite a number of empirically determined features of the symmetry detecting mechanism which will give us some basis for putting forward a neuro-model. I think this is encouraging because detecting symmetry is something that involves the whole of the pattern, and it is remarkable that one can do it with a high efficiency.

The figure of 50% is actually as high an efficiency as I have measured for any task, even the very simplest one of detecting the absence or presence of a weak light, at threshold, where it turns out that you only use about 50% of the information (Barlow, 1977). The figure of 50% is also reached, but not exceeded, when detecting variations of dot density (Barlow, 1978). To show I have no prejudices I also did this (with Aart van Meeteren) using sinusoidal modulations of dot density, and the answer to that is the same: provided that you don't have more than about 4 or 5 bars in the grating you perform with up to 50% efficiency. So apparently one can do this global type of task as efficiently as much simpler ones. One can visualize detecting local variation of dot density with local feature detectors and I like to think that the global task of symmetry detecting can also be done by a global feature detector along the general lines of Fig. 10, though we know that needs modifying.

I should mention a final conclusion from the high efficiencies of performance of these perceptual tasks, even though it is not so relevant to the theme of this conference. It is that the statistical limits of induction may be as important a factor for some higher perceptual tasks as physical factors, such as the quality of the image and the absorption spectra of pigments, are for simpler sensory tasks.

The Future of Feature Detectors

The first point to make is that those who explore the properties of single neurones in new parts of sensory systems, or well-known parts in new animals, are likely to need the concept of feature detectors in the future, as in the past, simply as a convenient qualitative description of experimental findings. The difficulty of discovering the specific stimulus that a particular cell demands has not been experienced by many of those who criticize the concept, and it is doubtful if they would be so critical if they had experienced it. For myself, the drama of resolving this difficulty was greatest when working with Levick and Hill on the rarer types of unit near the visual streak of the rabbit retina. One would know that the electrode was near a ganglion cell from the occasional, accidentally elicited, action potential, and one would know whereabouts in the visual field the receptive field lay from the postion of the electrode on the retina and from the previously plotted fields; but it might take an hour or more of frustrating trials before one discovered the stimulus that gave reliable, vigorous responses, and it was because of the drama of ultimately firing the cell reliably that we coined the phrase "trigger feature" for the appropriate patterned stimulus. Of course not all units are hard to trigger; some respond to an extended stimulus such as a line or edge of any orientation, and these are easy to track down. But if they require slow movement, or small size, or critical orientation, a rapid search is impossible.

The terms trigger-feature and feature-detector are useful for the qualitative description of units which are highly selective, though more quantitative descriptions are desirable and will, I have no doubt, come to be used more and more. However I personally shall continue to find a reliable qualitative description of a new type of selectivity more important and interesting than the quantification of established types.

Could one find feature detectors by other means than single-unit recording? The development of activity-sensitive anatomical methods such as the de-oxy-glucose technique raises fascinating possibilities, but of course the time scale of exploration is likely to make the discovery of new trigger-features impossible; one would never have thought of examining the cortex with this technique following uniocular exposure, or following exposure to striped fields (Hubel,Wiesel and Stryker, 1978) without the previous single unit results.

I mentioned at the beginning of this talk that the notion of "releasers", derived from ethological behavioural studies, had given an initial impetus to the discovery of feature detectors, and perhaps this source of discovery may not be exhausted. Those types of sensory stimulus that elicit distinct behavioural effects are also surely more likely to give definite neural responses. For example a combination of the ethological and the histological approaches might enable one to stain those cells in the visual pathway that are activated when a chick responds to a stimulus to which it is imprinted. At early levels activity can presumably only be triggered by stimuli for which the chick is endowed with feature-detectors, while at a higher level still activity should be aroused only by imprinted trigger features or combinations of trigger features. This is no more than fantasy, perhaps, but it is certainly stupid for physiologists and anatomists to neglect the possibilities suggested by behavioural studies.

Thus I think feature detectors definitely have a future for experimentalists, but is the concept also useful for those trying to understand and theorise about higher nervous function? Here again I think the answer is yes, and I will make this point in the form of the following, possibly controversial, postulate: The sensory performance of the whole animal is never better than that of at least one of its feature-detecting single units. The word "better" above is too broad, because the whole animal can certainly respond in a more versatile way than any one of its feature detectors, and it should be understood to mean "more efficient" in the statistical sense already explained. What is intended is the claim that if an animal reliably reacts differently in two situations, somewhere in its brain there must be at least one nerve cell that reliably responds differently in the two situations. The resonse is not necessarily always based on the same neurone, but the response of the whole animal to the whole set of situations can be no more reliable than the response of each single neurone to the subset of stimulations that it discriminates.

Clearly the future here lies with computer modelling, and I would be quite happy to see the term "feature detector" replaced by various types of linear operations and categorical descriptive terms in these representational models.

References

Adrian, E.D. (1947) The physical background of perception. Clarendon Press, Oxford.

Barlow, H.B. (1953) Summation and inhibition in the frog's retina. J. Physiol. 119, 69-88.

Barlow, H.B. (1977) Retinal and central factors in human vision limited by noise. In: Vertebrate Photoreception (Eds Barlow, H.B. and Fatt, P.) Ch. 19, pp 337-358. Academic Press: London

Barlow, H.B. (1978) The efficiency of detecting changes of density in random dot patterns. Vision Research 18, 637-650.

Barlow, H.B., Hill, R.M. and Levick, W.R. (1964) Retinal ganglion cells responding selectively to direction and speed of image motion in the rabbit. J. Physiol.,173, 377-407.

Barlow, H.B., Blakemore, C. and Pettigrew, J.D. (1967) The neural mechanism of binocular depth discrimination. J. Physiol. 193, 327-342.

Barlow, H.B. and Reeves, B.C. (1979) The versatility and absolute efficiency of detecting mirror symmetry in random dot displays. Vision Research, 19, 783-793.

Fulton, J.F. (1949) Physiology of the nervous system (3rd edition). Oxford University Press: New York.

Hartline, H.K. (1938) The response of single optic nerve fibres of the vertebrate eye to illumination of the retina. Am. J. Physiol. 121, 400-415.

Hartline, H.K. (1940b) The effects of spatial summation in the retina on the excitation of the fibers of the optic nerve. Am. J. Physiol. 130, 700-711.

Hubel, D.H. and Wiesel, T.N. (1959) Receptive fields of single neurones in the cat's striate cortex. J. Physiol. 148, 574-591.

Hubel, D.H. and Wiesel, T.N. (1962) Receptive fields, binocular interaction and functional architecture in the cat's visual cortex. J. Physiol. 160, 106-154.

Hubel, D.H., Wiesel, T.N. and Stryker, M.P. (1978) Anatomical demonstration of orientation columns in Macaque monkey. J. comp. Neirol. 177, 361-379.

Ingle, D. (1968) Visual releasers of prey-catching behavior in frogs and toads. Brain, Behaviour and Evolution 1, 500-518.

Lettvin, J.Y., Maturana, H.R., McCulloch, W.S. and Pitts,W.H. (1959) What the frog's eye tells the frog's brain. Proc. Inst. Rad. Engrg. 47, 1940-1051.

Levick, W.R. (1967) Receptive fields and trigger features of ganglion cells in the visual streak of the rabbit's retina. J Physil. 188, 285-305.

Lorenz, K.Z. (1952) King Solomon's Ring. Crowell.

Marshall, W.H., Woolsey, C.N. and Bond, P. (1941) Observations on cortical somatic sensory mechanisms of cat and monkey. J. Neurophysiol. 4, 1-24.

Maturana, H.R. and Frenk, S. (1963) Directional movement and horizontal detectors in pigeon retina. Science 142, 977-979.

Rosenblatt, F. (1959) Two theorems of statistical separability in the perception. Proceedings of a Symposium on the Mechanization of Thought Processes. pp 421-456. HMSO: London.

Sakitt, B. (1972) Counting every quantum. J. Physiol. 223, 131-150.

Selfridge, O.G. (1959) Pandemonium: a paradigm for learning. Proceedings of a Symposium on the Mechanization of Thought Processes held at the National Physical Laboratory. HMSO: London.

Tinbergen, N. (1951) The Study of Instinct. Oxford: Clarendon Press.

Waterman, T.H. and Wiersma, C.A.G. (1963) Electrical responses in decorpod crustacean visual systems. J. Cell. Comp. Physiol. 61, 1-16.

Waterman, T.H., Wiersma, C.A.G. and Bush, B.M.H. (1964) Afferent visual responses in the optic nerve of the crab Podophthalmus. J.Cell. Comp. Physiol. 63, 135-155.

THE ROLE OF TERMINATORS IN PREATTENTIVE PERCEPTION
OF LINE TEXTURES

Bela Julesz

Bell Laboratories
Murray Hill, New Jersey 07974

ABSTRACT

Recently, random-dot textures were found that in spite of their identical second-order statistics (hence power spectra) could still be effortlessly discriminated, based on conspicuous local features which include quasi-collinearity, corner, closure, and connectivity. Here, it will be shown, using textures composed of line segments, that these local features are not independent of each other, but can be described by two elementary units: <u>line segments</u> and their <u>terminators</u>. Furthermore, the preattentive texture perception system can count the number of terminators but ignores their positions. The line segments are a special case of <u>bars</u>, which were identically stimulated in these experiments. Besides color, these bars (of given orientation, width and length), their terminators (and crossed bars, as special case of terminators) are the elements of preattentive vision, to be called <u>textons</u>. In preattentive texture discrimination only the first-order statistics of these textons has perceptual significance, while the relative positions between the textons are unnoticed.

0. Prologue

It is my policy to decline invitations to symposia whose organizers insist on written contributions, since there is no excuse for cluttering the literature with already published findings, while a symposium proceeding is not the proper medium for entirely new results, for obvious reasons. Nevertheless, in the spring of 1979 I made an exception mainly because I had obtained some surprising new results in preattentive texture perception that I wanted to share and discuss with the unusually illustrious rostrum of participants of this Symposium. Now, more than two years later, reading my original manuscript sent me for final editing, I realize in what a dilemma I had gotten myself. During the last two years some further important developments took place in my laboratory that led to a reevaluation of my thinking and to the formulation of a novel "texton" theory of preattentive texture perception. These new results and insights were reviewed in a recent article of mine (Julesz, 1981).

So, the question arises what should I do with the original manuscript. The findings reported in it are still valid, and led in an important way to my current texton theory. On the other hand, the logic of building up the material is now straightforward, while the 1979 version had some important holes. Since some of the material in this article has not yet been published, I decided to stick as much to the original manuscript as I could, keep all the findings and demonstrations, but alter somewhat the logical flow. If the interested reader would take the trouble, and after reading this article would also consult my recent review article (Julesz, 1981), I hope the two together will yield a better understanding than each article by itself would.

1. Introduction

In 1978, three articles were published on random-dot texture perception in which this author and his co-workers reported the discovery of some local nonlinear features that seem to underlie preattentive perception (Caelli and Julesz, 1978; Caelli, Julesz and Gilbert, 1978; Julesz, Gilbert and Victor, 1978). These nonlinear features were found in textures with identical second-order statistics (i.e., with identical autocorrelations and, hence, with identical power spectra). It has been shown that a large class of iso-second order texture pairs exist, with different third- and higher-order statistics that cannot be effortlessly discriminated during a brief flash (under 200 msec duration) to prevent scrutiny, (Julesz, 1962, 1975; Julesz et al. 1973, Pratt et al. 1978). These findings led to the modified conjecture of the author (Julesz, 1981) that the preattentive texture discrimination system cannot process globally statistics of third- and higher-order. As a matter of fact, it is even doubted that globally the second-order statistics of textures could be evaluated by the preattentive visual system (Julesz, 1981). So, if one were to discover iso-second-order texture pairs that could be discriminated without scrutiny, this discrimination would have to be based on conspicuous local features. These putative local elements were named "textons" (Julesz, 1980, 1981), and the first such texton of quasi-collinear four disks was found by Caelli and Julesz (1978), as shown in Fig. 1A. In 1978 several other local conspicuous

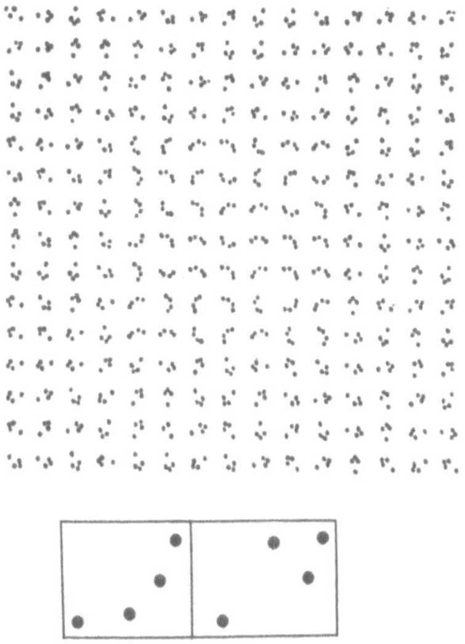

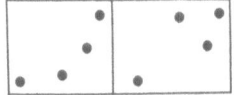

Fig. 1A

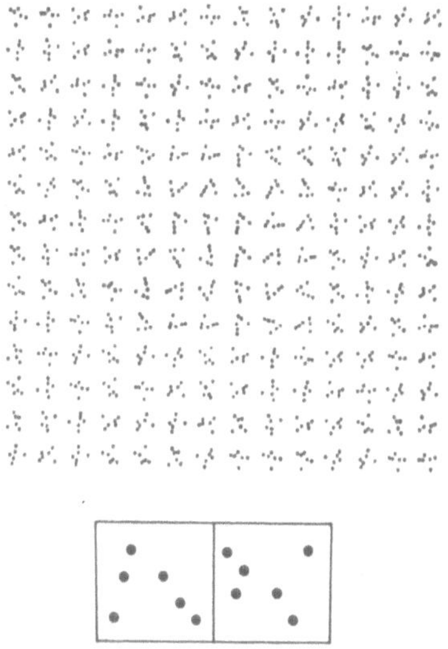

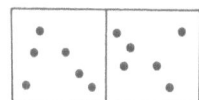

Fig. 1B

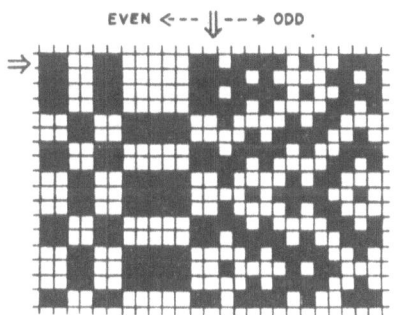

Fig. 1D

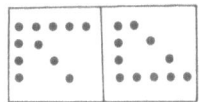

Fig. 1C

Figure 1: Discriminable texture pairs; counterexamples to the iso-power spectra texture conjecture. Discrimination is based on nonlinear local features of A) quasi-collinearity; B) corner; C) closure; and D) granularity (blobs). From Caelli et al., 1978, Julesz et al., 1978.

features were found in iso-second-order textures (Caelli, Julesz and Gilbert, 1978; Julesz, Gilbert and Victor, 1978; Victor and Brodie, 1978), and some are shown in Figs. 1B,C, and D. However, these features of corner, closure, granularity, and curvature are not separate textons. It will be shown that only underlined elongated blobs (particularly line segments) and their terminators (ends-of-lines) are textons, and corner, closure, granularity, and curvature can be explained by differences in width, length and orientation of the elongated blobs and in the number of their terminators. In addition to these textons, color is a third texton class. The terminator textons can be further subdivided into crossings (as shown by Bergen and Julesz, in preparation).

What is even more interesting, the preattentive visual system seems to ignore the relative position (phase) between textons. Only differences in the first-order statistics of textons have perceptual significance.

Interestingly (although not unexpectedly) these "perceptual elements" of texture discrimination - identified by psychological methods - are similar to some of the neurophysiological feature extractors found in the cortex of the monkey (Hubel and Wiesel, 1968). Particularly, feature extractors of "line segments", "bars", and "edges" (tuned to specific orientation, width, aspect ratio, area, etc.), and of "terminators" (tuned to the end of lines, or boundaries of certain domains) were suggested by the simple, complex and hyper-complex cortical units of the neurophysiologists.

In this article only textures composed of line segments are studied, and it is shown how "perceptual elements" of closure, connectivity, corner, etc. discovered in random dot textures can be described by some even more primitive elements, the textons such as "line segments" and their "terminators".

2. Experimental Procedure

The line textures used in these experiments were generated by a PDP 11/20 computer and displayed by an HP electrostatically deflected monitor (with P4 fast phosphor), as described in detail elsewhere (Caelli and Julesz, 1978). The display format used on the HP monitor will be discussed in the next section. However, the demonstrated figures reproduced here were plotted by a Tektronix video display system. Each texture pair was constructed as a 14x14 array of micropatterns. All micropatterns within a 6x6 subarray set in one quadrant of the larger array were of one type, and all remaining micropatterns were of the second type. The reader's task is to discriminate the texture regions containing these two micropattern types, and identify the quadrant in which the subarray occurred. For most micropattern pairs used here (since they were confined within a rectangular lattice) the textures could be constructed with equivalent second-order statistics by throwing the micropatterns in eight specific ways: rotated by 0, 90, 180, and 270 deg arc, either in their original form or by taking their mirror image. The micropatterns did not overlap, and their size was identical, but could be zoomed to any size by a single instruction. Also, the micropatterns could be centered on the regular lattice points, or be jittered around these points according to a chosen probability distribution. In several cases texture discrimination ability depended on which one of the textures was placed inside or outside. The introduction of jitter did not appreciably affect discriminability, provided the micropattern size was small compared to their average separation. So, except for Figs. 6F and 6G where the micropatterns are jittered, the textures shown are regularly spaced.

The reader should be able to verify the reported findings with the demonstration figures in this article, if he refrains from scrutinizing the texture elements. A simple device may ensure that only preattentive processes are utilized, if the reader has access to a shutter (i.e., a photographic camera with removable lens and back plate): he can inspect the texture pair through this shutter for 1/5 second or briefer duration, thus preventing scanning eye-movements or shifts of attention.

3. Experimental Results

Observer sat at 110 cm distance from the Hewlett-Packard screen on which the texture array was flashed for various short time intervals. The array extended 14 cm x 14 cm, so the viewing angle was 7.28 deg x 7.28 deg, and thus each micropattern was confined in a 73 min arc x 73 min arc area. This micropattern separation (L) was large enough to prevent adjacent micropatterns from getting too close together as they were increased in size to obtain adequate resolution. Indeed, the

micropattern size was selected such that when fixating in the center
of the 6Lx6L array, the micropatterns in the farthest corner could be
resolved (when presented in isolation for a brief flash) as shown by
the dotted circles in Fig. 2A.

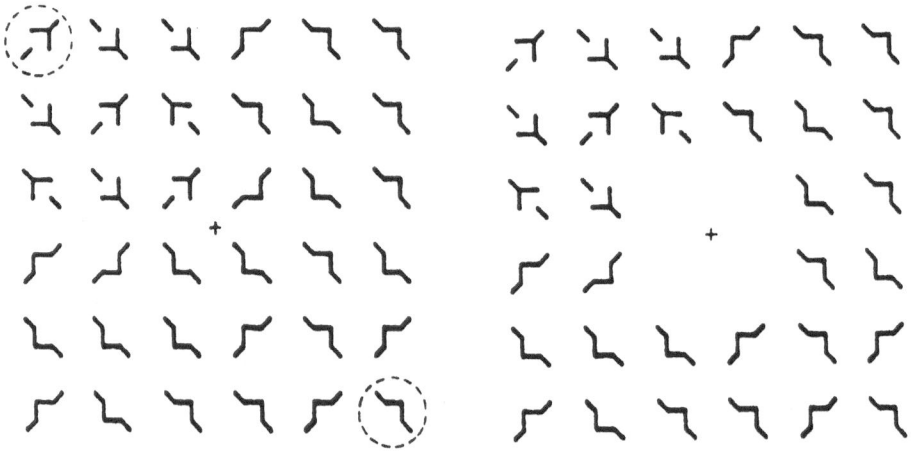

Fig. 2A   Fig. 2B

Figure 2: Experimental   procedure   for   texture
discrimination. a) Indicates how the size of individual
micropatterns in the furthest two corners is selected to
yield 80% correct discrimination, b) The actual array shown,
with center blanked out to prevent foveal scrutiny.

Furthermore, the 73 min arc micropattern separation was somewhat
larger than the radius of foveal attention around the center of gaze.
In order to prevent observers from scrutinizing the texture pair by
shifts of foveal attention, a 2Lx2L area in the center of the 6Lx6L

array was kept blank as shown in Fig. 2B. Thus one element was left
blank within the 3Lx3L smaller texture array which appeared in one
quadrant of the 6Lx6L texture array. The micropatterns for the
texture pairs were randomly thrown in 16 ways of equal probability: 0,
45, 90, 135, 180, 225, 270, and 315 deg arc, either in their original
form or by taking their mirror images. Observer indicated his
response - whether the inner texture was in the upper-left, upper-
right, lower-left, or lower-right quadrant - by pressing one of the
keys (1, 2, 3, or 4, respectively). This was accompanied by a tone
signal when a mistake was committed. Only after this forced choice
response was completed, could the observer obtain a new stimulus, by
pressing the start key.

Since the task was forced choice between four quadrants, chance
performance (i.e., nondiscrimination) was 25% correct responses, while
ideal performance was 100% correct responses. Since the task was
difficult for weakly discriminable textures, the best observers had
practice sessions before the two trial sessions. All sessions
consisted of 40 trials with 50% probability that each texture trial
formed the inner or outer textures. Discrimination scores, expressed
as the percent correct responses are shown for our best observers (PR)
and (PB) in Table 1. Each row corresponds to one of the texture pairs
demonstrated in this article, giving their actual figure number. Each
correct response in percent thus corresponds to a given texture pair
tested 80 times for a given T flash duration given in parenthesis.
This author also served as an observer, and his scores were similar
but never better than those of PB.

The portraying time (T) of the textures was usually kept at 200
msec or shorter, and Table 1 shows the correct score of two observers
for these durations. However, T could be considerably increased
without affecting discrimination scores. For instance, the
nondiscriminable texture pairs of Fig. 7 yielded a 25% chance score
for (PR) and 29% for (PB), which remained the same for T=352 msec.
Only around T=440 msec did (PB)'s score increase to 35%, indicating
that observer was able to perform two saccadic eye movements, which
permitted him to learn some strategy of scrutinizing the stimulus. In
contrast, the texture pair of Fig. 8 for T=200 msec was discriminable
yielding 72% correct score for (PR) and 57% for (PB). The strongly
discriminable texture pair of Fig. 6 yielded a 90% score for (PR) and
95% for (PB), at T=200 msec.

			(PR)	(PB)
Fig. 6			87%	90%
Fig. 7			24%	29%
Fig. 8			36%	40%
Fig. 9			70%	60%
			77%	64%
Fig. 10			97%	100%

Table 1: Correct scores of texture discrimination in 80 trials by two observers, for T<200 msec, 6x6 micropattern large texture array with a 3x3 small array in one quadrant, minus a 2x2 center array kept blank around fixation point. Figure numbers (and micropattern duals) correspond to those in the article (the latter consisting of a 14x14 large array, and a 6x6 small array).

One can simplify the data of Table 1 by dividing texture discrimination into two categories: a) nondiscriminable, if at T<200 (or, perhaps 300) msec the correct scores are about 25%, b) discriminable, otherwise. One might talk also about strong and weak discrimination. In this article, 50% or larger correct scores at T<200 msec will be regarded as strong texture discrimination, while under 50% correct scores at T<200 msec will be regarded as weakly discriminable, provided they are significantly higher than 25%. Of course, all these measures require highly trained observers. And yet, it is interesting that the data of Table 1 will describe in most cases the direct impressions of the reader as he inspects the following demonstrations, and refrains from scrutiny.

4. Iso-Second-Order Line Textures

In the introduction it was emphasized that a large class of random-
dot textures with identical second-order statistics does exist that
cannot be discriminated without scrutiny. Since these
indistinguishable texture pairs differ in their third- and higher-
order statistics, therefore in preattentive texture perception third-
or higher-order statistical parameters are not processed. This global
limitation on computational power by the preattentive system is so
important that we demonstrate a few such examples.

The first method, using mirror image dual micropatterns, was
devised by Julesz, et al. (1973), and is illustrated by Fig. 3.
Figure 3A depicts the two micropatterns, an "R" and its mirror image
dual, out of which the two textures are generated, respectively. In
Fig. 3B the internal area consists of the "R" shaped micropatterns
while in Fig. 3C they constitute the surround. Both the inside and
outside textures consist of micropatterns selected at random
orientations (at multiples of 90 deg arc). An examination of Figs.
3B and 3C will show that discrimination is impossible without
scrutiny. As of now, no mirror image texture duals have been found
that can be discriminated during a brief flash under 200 msec. In
contrast to other iso-dipole construction methods, the mirror image
method generates texture pairs consisting of the same line segments
with identical length and orientation, and identical number of end-of-
lines (terminators), although the phase (position) spectra are
different. This experiment clearly supports the conjecture that in
global texture perception the local position of line segments and
their terminators is lost, provided the "perceptual elements" that are
formed by these line segments are the same, and occur with equal
frequency, in both textures.

The second demonstration, based on the generalization of the four-
disk method (see Fig. 9) by Caelli, Julesz, and Gilbert (1978) is
shown in Fig. 4. [This is essentially the same as Fig. 11 of Caelli
et al., (1978) except that in place of circles rectangles are used
here.] Figure 4A shows the two micropatterns used to form the two
textures of Figs. 4B and 4C, which have iso-power-spectra. Again,
Fig. 4B is the same as Fig. 4C, except the inside-outside textures
are exchanged. Discrimination of these textures must be based on
positional information: whether the two dots fall in one rectangle or
in two. Again it seems that this information is lost in preattentive

Fig. 3A

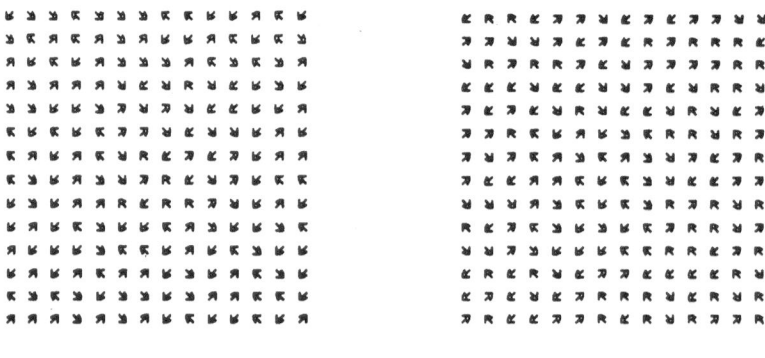

Fig. 3B   Fig. 3C

Figure 3: Nondiscriminable iso-power spectrum texture pair
generated by the mirror-image dual method of Julesz et al.
(1973).

perception. The very regular "crystalline" form of the textures in
Figs. 4B and 4C could possibly overpower the fine details, such as
the distribution of dots. A jitter may be introduced to break up this
lattice structure as shown in Figs. 4D and 4E. Discrimination is
still impossible.

The demonstration of Fig. 4 contained both dots and line segments.
Since in this study the emphasis is on line segments, in Fig. 5 the
same iso-dipole texture pair is generated, except that x-shaped line

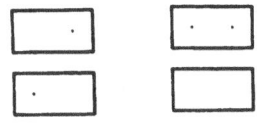

Fig. 4A

Fig. 4B

Fig. 4C

Fig. 4D

Fig. 4E

Figure 4: Nondiscriminable iso-power spectrum texture pair, demonstrating the insensitivity of texture perception to phase (position) information.

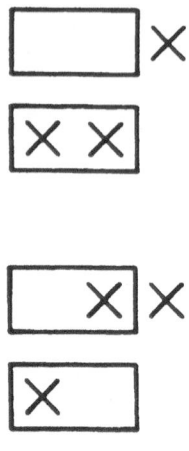

Fig. 5A

Fig. 5B

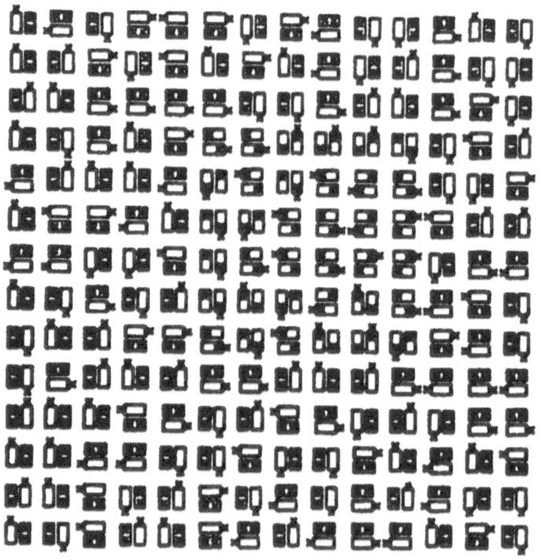

Fig. 5C

Figure 5: Similar to Fig. 2, except that for dots x-shaped elements are used.

segments are used in place of dots. As Figs. 5B and 5C demonstrate, texture discrimination is very weak, if not impossible.

5. Closures, a Special Case of Connectivity

The next demonstration, using the same method that was applied to generate Figs. 4 and 5, is depicted by Fig. 6A. One micropattern is connected, while its dual is not, and the resulting texture pairs of Figs. 6B and 6C are easily discriminated, even though the textures have identical power spectra. [It appears to the author that the connected micropattern texture is easier to detect when inside (Fig. 6B), while the boundaries between textures are easier to trace when the nonconnected micropattern texture is inside (Fig. 6C).] Is discrimination on the basis of connectivity making this another "perceptual element", or is it simply a change in the number of terminators? Perhaps the strong discrimination of iso-dipole textures composed of closed versus open micropatterns in the demonstration of Caelli, et al. (1978), is not on the basis of closure, but again reflects changes in the number of terminators?

That, indeed, this is the case is shown by Fig. 7, in which "closure" and "connectivity" are juxtaposed. Both textures are composed of micropatterns with two terminators each, and cannot be discriminated without scrutiny. It should be stressed that the texture pairs of Figs. 7B and 7C are not quite iso-second-order, hence have different power spectra and yet are not discriminable. Many such examples were known of textures with different power spectra that did not yield discrimination (Julesz, 1962, 1975; Julesz et al., 1973; Mayhew and Frisby, 1978). This means that the preattentive perceptual system does not utilize the entire second-order statistics, but only some subset of it. As a matter of fact, it was recently shown (see Fig. 7 in Julesz, 1981) that without first-order differences in textons, large differences in the second-order statistics are indistinguishable.

6. The Role of Terminators in Texture Discrimination

The inability to discriminate between the connected (but open) and closed (but unconnected) textures in Figs. 7B and 7C might be the result of these elongated micropatterns having their terminators outside some "local area of statistical evaluation". Indeed, Stevens

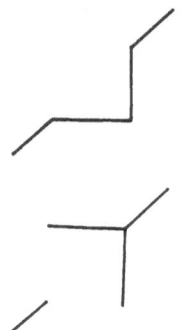

Fig. 6A

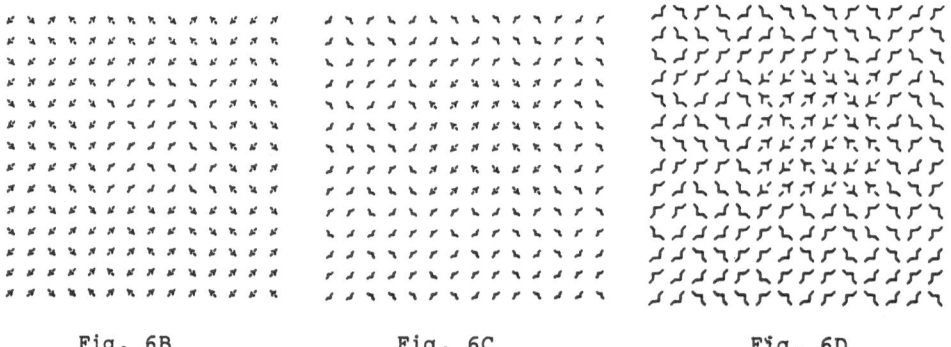

Fig. 6B   Fig. 6C   Fig. 6D

Fig. 6E 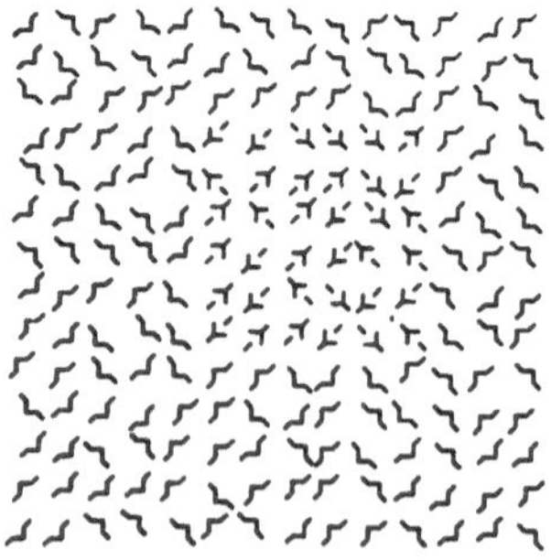   Fig. 6F    Fig. 6G

<u>Figure 6</u>: Discriminable iso-power-spectra texture based on connectivity.

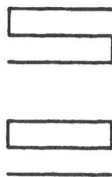

Fig. 7A

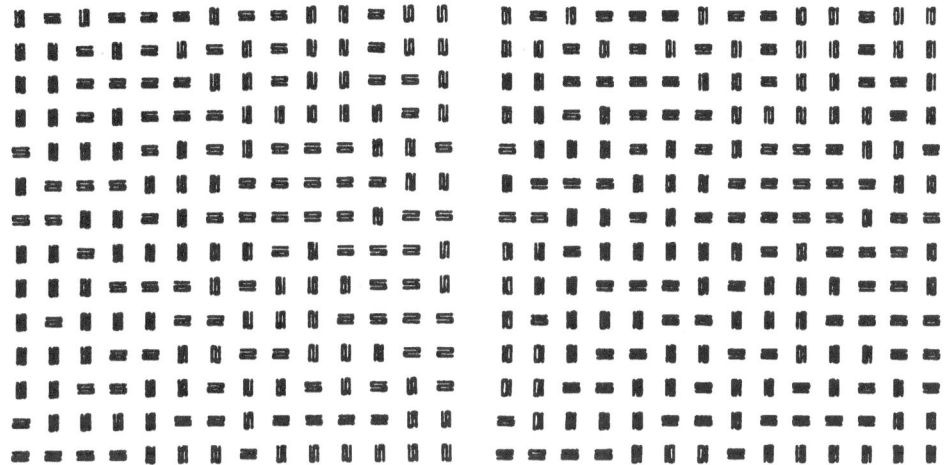

Fig. 7B   Fig. 7C

<u>Figure 7</u>: Demonstration that textures composed of connected
(open) and unconnected (closed) micropatterns, respectively,
cannot be effortlessly discriminated, if the number of their
<u>terminators</u> agrees. Power spectra are slightly different.

(1978) studying the perception of Glass-patterns postulated such areas
of limited size.

That this is not the case is shown in Fig. 8. Here the same
micropattern pair is used as in Fig. 8, except that their extent is
confined within a much smaller area. Since Figs. 7 and 8 are not
iso-dipole, the shorter dipole differences in Fig. 8 are more
conspicuous than the longer dipole differences in Fig. 7, a fact
already known (Julesz, 1975; Caelli and Julesz, 1978). Texture

Fig. 8A

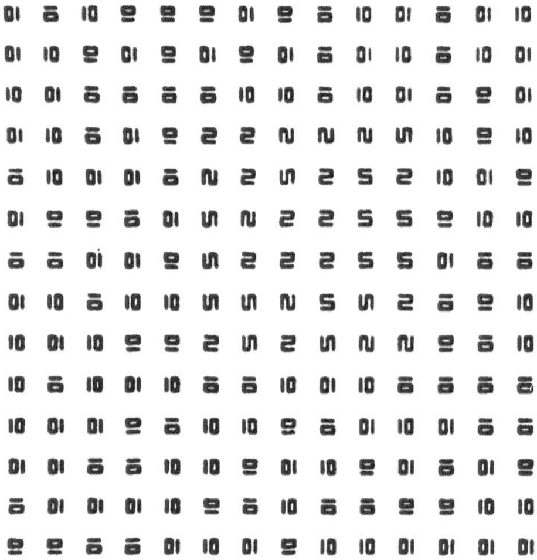

Fig. 8B

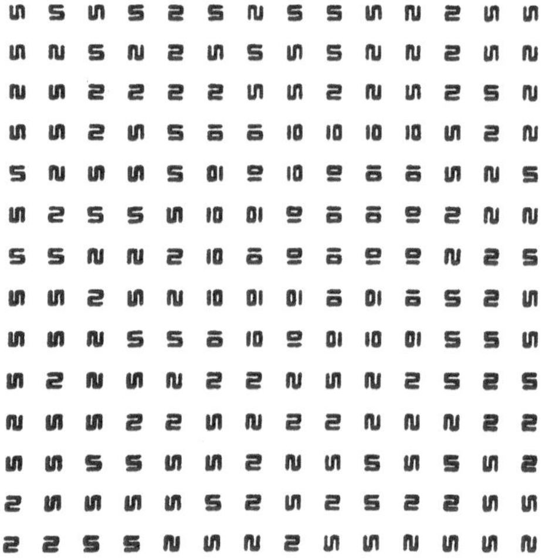

Fig. 8C

Figure 8: Similar to Fig. 7, except that the micropattern
duals are more compact. Power spectra are greatly
different.

discrimination in Fig. 8 appears stronger than it is for Fig. 7, in
a flash of 200 ms or briefer duration. (PR)'s score is 37%, while
only 24% for Fig. 7.

It should be stressed that the elongated blob textons can be "on"
or "off" types. The white gap in the "open" micropatterns is more
conspicuous in Fig. 8 than in Fig. 7 which might easily account for
the discrimination in Fig. 8.

A somewhat stronger discrimination is also expected when one of the micropatterns has two terminators, while its micropattern partner has three. This is shown in Fig. 9.

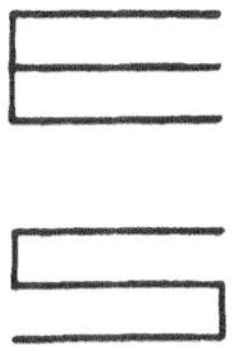

Fig. 9A

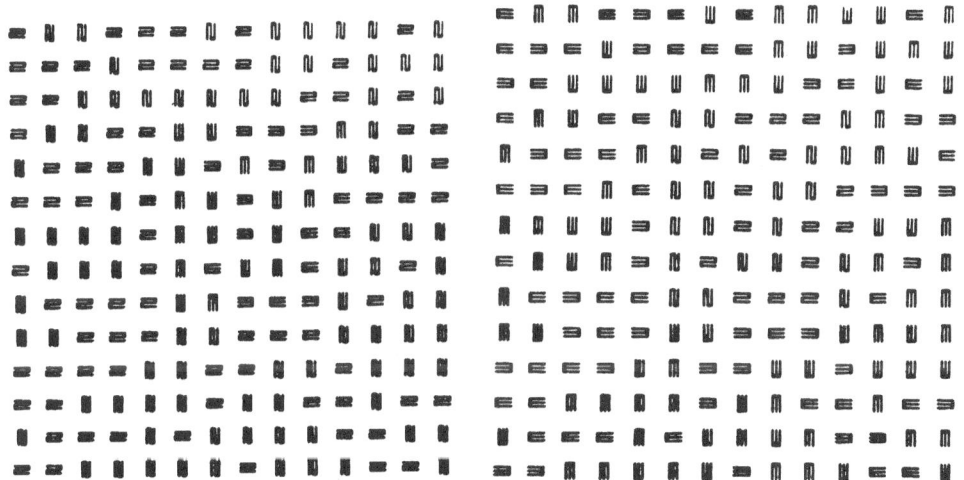

Fig. 9B   Fig. 9C

Figure 9: Discriminable texture pair based on the difference in the number of terminators of the micropatterns. Power spectra are slightly different.

Indeed, there is a 72% score, much better than chance for T=195ms. While the second-order statistics are different for textures in Figs. 9B and 9C, their first order statistics agree. If these micropatterns are shortened (as in Fig. 8) observer's response is 72% correct.

Of course, if both the second- and first-order statistics differ,

and there is also a large difference in the number of terminators,
texture discrimination becomes strong, as shown in Fig. 10. (80%
correct discrimination can be obtained in a 165 msec flash, as shown
in Table 1.) This example is not very interesting, except that it
shows that the "local area of statistical evaluation" in Fig. 7 is
not too elongated for computing various statistics, since the
micropatterns in Fig. 10 are similarly elongated.

Fig. 10B   Fig. 10C

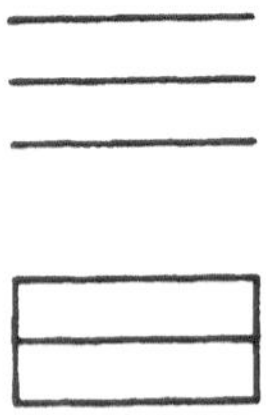

Fig. 10A

Figure 10: Discriminable texture pair with different first-
order statistics, to demonstrate that terminators of the
similarly elongated micropatterns in Fig. 7 can be resolved
perceptually.

From these experiments we can conclude that the failure to

discriminate between the connected and closed textures in Fig. 7 is due to the fact that their micropatterns have the same line segments and the same number (2) terminators. It seems that the preattentive perceptual system <u>can</u> <u>count</u> <u>only</u> <u>the</u> <u>number</u> <u>of</u> <u>terminators,</u> <u>but</u> <u>ignores</u> <u>their</u> <u>positions.</u>

7. The Role of Dots in Connectivity

Finally, we turn to the question: how quasi-collinear dots interact with line segments, using connectivity as a criterion. Let us take Fig. 6A, but place two dots at "strategic" positions as shown by Fig. 11A. These iso-power-spectra textures are now rendered not discriminable, as demonstrated by Figs. 11B and 11C. It appears that the two dots and the two corner points of the line segments formed a collinear structure that connected as strongly as an actual line segment. In Fig. 6A the connected micropattern had two terminators, while the unconnected micropattern had four, resulting in strong discrimination. In Fig. 12A the number of terminators became the same for both micropatterns. On the other hand, if the two dots are placed elsewhere, as shown in Fig. 12A, so that the corner points of the line segments are not collinear with the two dots, strong discrimination is experienced in Figs. 12B and 12C. When dots become "connectors" they cease to act as terminators. A more detailed account on the interaction of dots with line segments in texture perception can be found elsewhere (Julesz, 1979).

8. Discussion

In the light of these experiments it seems that the local elements of quasi-collinearity, corner, closure, and "blobs" (which yielded discrimination in iso-second-order texture pairs) are not independent of each other. If we ignore "blobs" that might stimulate bar detectors, and restrict ourselves to line textures composed of thin line segments, then all "perceptual elements" can be further reduced to more elementary units, the <u>textons</u> of <u>line</u> <u>segments</u> and their <u>terminators.</u> The quasi-collinear feature is identical to a line segment. The <u>corner</u> between two line segments ending at the crossing point versus protruding through the crossing point yielded strongly discriminable textures (see Fig. 5 in Caelli et al., 1978). This discrimination might be the result of terminator difference. The true corner has only two terminators (perhaps three, if the corner point is also included), while the protruding corner has four terminators.

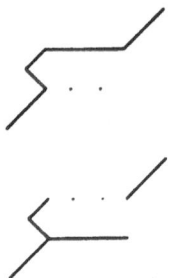

Fig. 11A

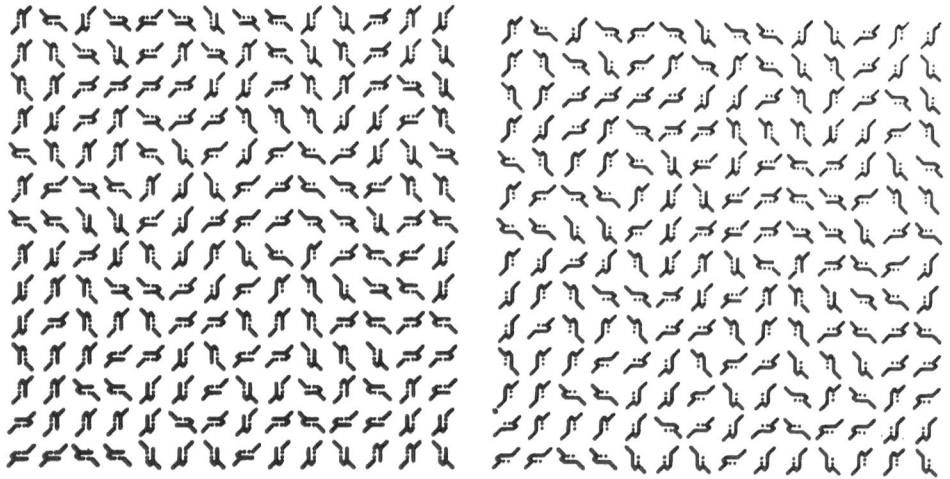

Fig. 11B   Fig. 11C

Figure 11: Demonstration of illusory (virtual) line
segments acting as real lines if dots and breaks in lines
are collinear.

However, it is possible to regard the crossing of two line segments
as a third texton class. Indeed, while it is impossible to
discriminate between texture pairs composed of T- and L-shaped
micropatterns (particularly if a small gap exists between the two
perpendicular line segments that form the T-s and L-s), very strong
discrimination is experienced between textures composed of + versus L
(or + versus T)-shaped elements (Bergen and Julesz, in preparation).

The closure feature, as we have seen, is a special case of

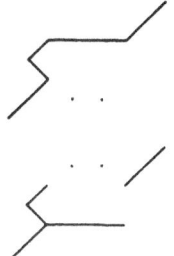

Fig. 12A

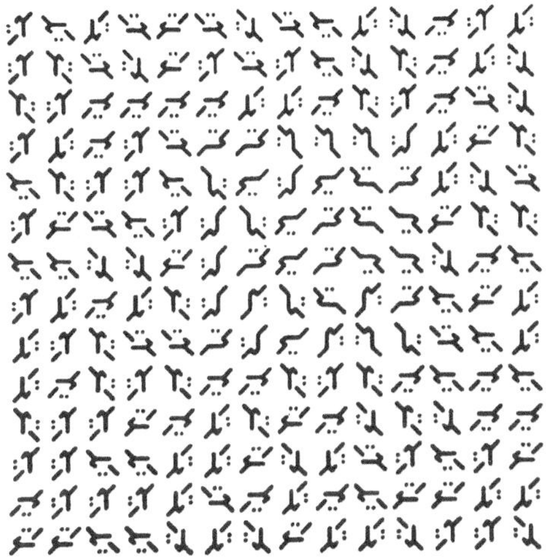

Fig. 12B 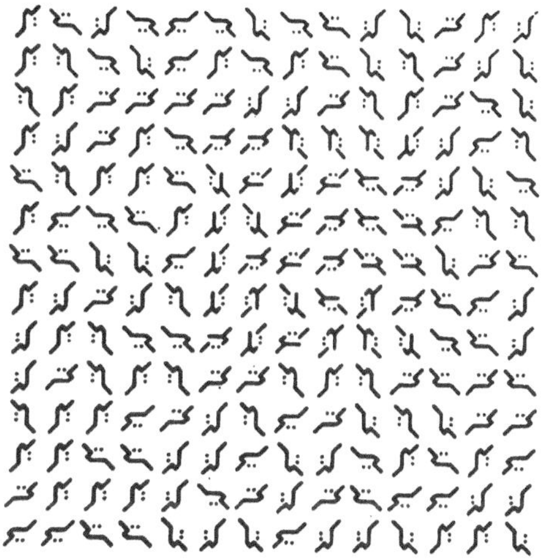   Fig. 12C

Figure 12: Demonstration of illusory (virtual) line segments remaining virtual lines if dots and breaks in lines are not collinear.

connectivity. The difference between a closed and open micropattern, or between a connected and unconnected one, can be simply explained by differences in the number of their terminators. For instance, the minimum number of line segments that can form a closed area is three. The closed triangle and its iso-dipole open dual (in Fig. 1C) have zero, and three terminators, respectively. This large terminator difference explains why this case yields the strongest discrimination among all discriminable iso-second-order textures of Caelli et al. (1978), as shown in Fig. 1C.

It seems that the perception of line textures is based on some statistics of line segments (with the same orientation and extent), and their terminators, but the exact position of these line segments and terminators is not utilized. Indeed, Julesz et al. (1973) stated: "It might be that the texture discrimination process takes only the first-order statistics of various simple feature extractors that might be segregated according to diameter (and for those with elongated receptive fields, according to width and orientation)." However, they did not know at the time the importance of terminators. It was David Marr (1976) who first emphasized the importance of terminators, besides line segments, in the early processing of visual information. His "primary sketch" model appears similar to the ideas developed here. Nevertheless, there is a crucial difference between Marr's approach and that of this author. Marr developed his model within the framework of artificial intelligence, trying to invent algorithms that are able to perform some well developed perceptual tasks, inspired by the feature extractors of single micro-electrode neurophysiology. On the other hand, the findings reported here were derived by strictly psychophysical methods, and these investigators were often skeptical of the role of the highly local neurophysiological feature extractors in global perceptual phenomena. But more importantly, Marr believed that "place tokens," higher-order constructs formed by line segments and terminators are essential in preattentive vision. He did not guess, however, that in preattentive texture discrimination no such place tokens are utilized, and only the first-order statistics of textons have perceptual significance. Similarly, it is most unlikely that the present enterprise of single microelectrode neurophysiology could have revealed that in the preattentive perceptual state the relative positions of textons (feature detectors) are ignored.

In spite of these caveats, the fact that the textons discovered by this author and his co-workers resemble some of the cortical feature extractors and some of the feature detectors that Marr invented, is most gratifying.

9. Conclusions

In this article the textures under study were composed of line segments, forming perceptual elements. It turned out that the first-order statistics of line segments and their terminators could explain the conspicuousness of perceptual elements, provided the position of these textons is ignored. The positional information of line segments is perhaps retained to some extent, after all changes in line segment configuration can alter the number of terminators, or yield crossings, and parallel line segments at various distances might stimulate "bar" detectors of specific widths.

The role of these bar detectors is beyond the scope of this study. It remains to be seen what the role of bar detectors is in preattentive texture perception, and what are the terminators of bars. Whether some simple statistics of specific line segments, bars, and their terminators are adequate to explain texture discrimination, or some further textons will be found is an interesting problem, worthy of study.

In summary, it has been shown that in preattentive texture perception only the first-order statistics of textons (e.g. their density) can be detected. Besides color, only elongated blobs (line segments), their terminators (ends-of-lines) and crossings of line segments are textons. Without scrutiny the relative position between line segments, their terminators and crossings cannot be perceived. In other words, in the preattentive perceptual state there is no "coupling" between the textons. Differences in texton density are instantaneously detected, and this parallel system of preattentive vision serves as an "early warning system" and alerts the serial system of focal attention to inspect the areas of texton density changes. Only in a minute area of focal attention does the coupling between textons exist, enabling us to tell, say, an "R" from its "mirror-image".

Whether these few texton classes are used only in preattentive texture perception, while in focal attention, such as in form recognition some other "elements of form" do exist, remains to be seen. Nevertheless, the finding that without some density changes in textons we have no initiative to shift our focal attention is of interest in understanding how figure is selected from the ground.

10. Appendix

In Fig. 13 it is shown to generate iso-power spectra textures by extending the four-disk method of Julesz et al. (1973), a novel approach invented by Caelli, Julesz, and Gilbert (1978). All the iso-dipole (iso-power-spectra) textures demonstrated here were generated by this method, except the texture pair of Fig. 1D, that is not only iso-dipole but even iso-trigon, and was invented by Julesz, Gilbert

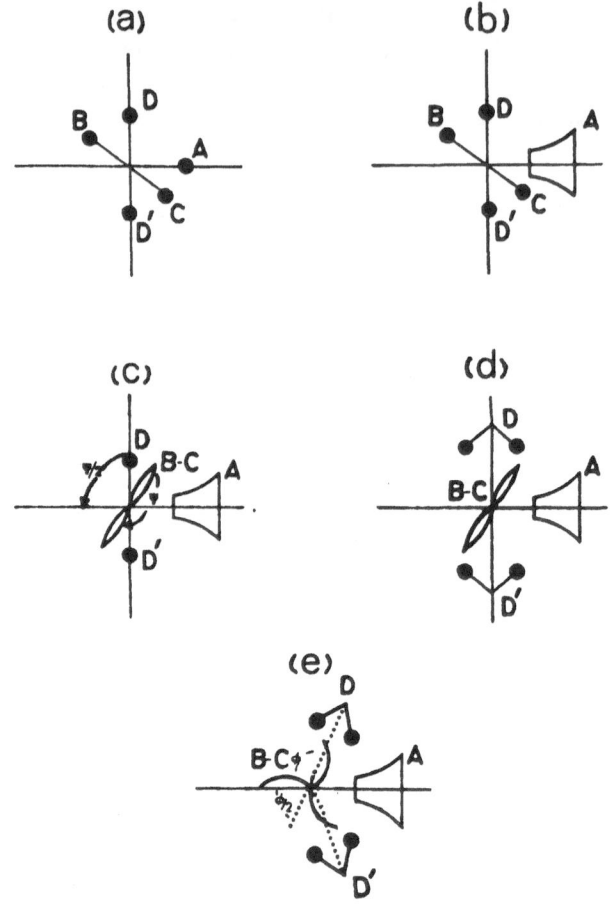

Fig. 13A-E

Figure 13: Methods to generate iso-power spectra textures (e.g., Figs. 4 and 7). From Caelli, Julesz and Gilbert (1978).

and Victor (1978).

11. Acknowledgments

The author thanks Dr. Peter Burt for the development of the display program that greatly facilitated texture generation. Special thanks to Prof. Werner Reichardt and Dr. Tomaso Poggio whose penetrating questions to this author on texture perception led to this article.

References

Barlow, H.B. 1978. The efficiency of detecting changes of density in random dot patterns. Vision Res. 18, 637-650.

Caelli, T.M. and Julesz, B. 1978. On perceptual analyzers underlying visual texture discrimination: Part I. Biol. Cybernetics, 28, 167-175.

Caelli, T.M., Julesz, B. and Gilbert, E.N. 1978. On perceptual analyzers underlying visual texture discrimination: Part III. Biol. Cybernetics, 29, 201, 214.

Caelli, T.M. and Julesz, B. 1979. Psychophysical evidence for global feature processing in visual texture perception. J. Opt. Soc. Am., 69, 675-678.

Campbell, F.W. and Robson, J.C. 1968. Application of Fourier analysis to the visibility of gratings. J. Physiol. Lond., 197, 551-566.

Enroth-Cugel, C. and Robson, J.G. 1966. The contrast sensitivity of retinal ganglion cells of the cat. J. Physiol. Lond., 187, 517-522.

Graham, N. and Nachmias, J. 1971. Detection of grating patterns containing two spatial frequencies: A comparison of single channel and multichannel models. Vision Res., 11, 251-259.

Hubel, D.H. and Wiesel, T.N. 1962. Receptive fields, binocular interaction and functional architecture in the cat's visual cortex. J. Physiol. Lond., 160, 106-154.

Hubel, D.H. and Wiesel, T.N. 1968. Ruceptive fields and functional architecture of monkey striate cortex. J. Physiol. Lond., 195, 215-243.

Julesz, B. 1962. Visual pattern discrimination. IRE Transaction of Information Theory IT-8, 84-92.

Julesz, B. 1965. Texture and visual perception. Sci. Am., 212, 38-48 (Feb. issue).

Julesz, B. 1971. Foundations of Cyclopean Perception. Chicago: The University of Chicago Press.

Julesz, B. 1975. Experiments in the visual perception of texture. Sci. Am., 232, 34-43 (Apr. issue).

Julesz, B. 1980. Spatial nonlinearities in the instantaneous perception of textures with identical power spectra. Phil. Trans. Roy. Soc., B 290, 83-94.

Julesz, B. 1981. Textons, the elements of texture perception, and their interactions. Nature, 290, 12 March, 91–97.

Julesz, B., Frisch, H.L., Gilbert, E.N. and Shepp, L.A. 1973. Inability of humans to discriminate between visual textures that agree in second-order statistics – revisited. Perception, 2, 391–405.

Julesz, B., Gilbert, E.N. and Victor, J.D. 1978. Visual discrimination of textures with identical third-order statistics. Biol. Cybernetics 31, 137–140.

Julesz, B. and Caelli, T.M. 1979. On the limits of Fourier decompositions in visual texture perception. Perception 8, 69–73.

Marr, D. 1976. Early processing of visual information. Phil. Trans. Roy. Soc. B 275, 483–524.

Mayhew, J.E.W. and Frisby, J.P. 1978. Texture discrimination and Fourier analysis in human vision. Nature 275, 438–439.

Pratt, W.K., Gaugeras, O.D. and Gagalowicz, A. 1978. Visual discrimination of stochastic texture fields. IEEE Trans. Syst. Man and Cybernetics, SMC-8 (11), 796–804.

Schatz, B. 1978. The computation of immediate texture discrimination. Computer Sci. Dept. Report 152, Carnegie-Mellon Univ., Pittsburgh, Pa.

Stevens, K.A. 1978. Computation of locally parallel structure. Biol. Cybernetics 29, 19–28.

Uttal, W.R. 1975. An autocorrelation theory of form detection. Hillsdale, New Jersey: Lawrence Erlbaum Associates.

Victor, J.D. and Brodie, S.E. 1978. Discriminable textures with identical Buffon-needle statistics. Biol. Cybernetics 32, 231–234.

VISUAL INFORMATION PROCESSING:

THE STRUCTURE AND CREATION OF VISUAL REPRESENTATIONS

David Marr

M.I.T. Artificial Intelligence Laboratory
and Department of Psychology
545 Technology Square, Cambridge, MA 02139

Summary

For human vision to be explained by a computational theory, the
first question is plain: What are the problems the brain solves when
we see? It is argued that vision is the construction of efficient
symbolic descriptions from images of the world. An important aspect
of vision is therefore the choice of representations for the different
kinds of information in a visual scene. An overall framework is
suggested for extracting shape information from images, in which the
analysis proceeds through three representations; (1) the primal
sketch, which makes explicit the intensity changes and local two-
dimensional geometry of an image, (2) the 2 1/2-D sketch, which is a
viewer-centred representation of the depth, orientation and
discontinuities of the visible surfaces, and (3) the 3-D model
representation, which allows an object-centred description of the
three-dimensional structure and organization of a viewed shape. The
critical act in formulating computational theories for processes
capable of constructing these representations is the discovery of
valid constraints on the way the world behaves, that provide
sufficient additional information to allow recovery of the desired
characteristic. Finally, once a computational theory for a process
has been formulated, algorithms for implementing it may be designed,
and their performance compared with that of the human visual
processor.

Introduction

Modern neurophysiology has learned much about the operation of the
individual nerve cell, but unpleasantly little about the meaning of
the circuits they compose in the brain. The reason for this can be
attributed, at least in part, to a failure to recognize what it means

to understand a complex information-processing system; for a complex system cannot be understood as a simple extrapolation from the properties of its elementary components. One does not formulate, for example, a description of thermodynamical effects using a large set of equations, one for each of the particles involved. One describes such effects at their own level, that of an enormous collection of particles, and tries to show that in principle, the microscopic and macroscopic descriptions are consistent with one another.

The core of the problem is that a system as complex as a nervous system or a developing embryo must be analyzed and understood at several different levels. Indeed, in a system that solves an information processing problem, we may distinguish four important levels of description (Marr and Poggio 1977, Marr 1977a). At the lowest, there is basic component and circuit analysis -- how do transistors (or neurons), diodes (or synapses) work? The second level is the study of particular mechanisms: adders, multipliers, and memories, these being assemblies made from basic components. The third level is that of the algorithm, the scheme for a computation; and the top level contains the theory of the computation. A theory of addition, for example, would encompass the meaning of that operation, quite independent of the representation of the numbers to be added -- say Arabic versus Roman. But it would also include the realization that the first of these representations is the more suitable of the two. An algorithm, on the other hand, is a particular method by which to add numbers. It therefore applies to a particular representation, since plainly an algorithm that adds Arabic numerals would be useless for Roman. At still a further level down, one comes upon a mechanism for addition -- say a pocket calculator -- which simply implements a particular algorithm. As a second example, take the case of Fourier analysis. Here the computational theory of the Fourier transform -- the decomposition of an arbitrary mathematical curve into a sum of sine waves of differing frequencies -- is well understood, and is expressed independently of the particular way in which it might be computed. One level down, there are several algorithms for computing a Fourier transform, among them the so-called Fast Fourier Transform (FFT), which comprises a sequence of mathematical operations, and the so-called spatial algorithm, a single, global operation that is based on the mechanisms of laser optics. All such algorithms produce the same result, so the choice of which one to use depends upon the particular mechanisms that are available. If one has fast digital memory, adders, and multipliers, one will use the FFT, and if one has a laser and photographic plates, one will use an "optical" method.

Now each of the four levels of description will have its place in the eventual understanding of perceptual information processing, and of course there are logical and causal relations among them. But the important point is that the four levels of description are only loosely related. Too often in attempts to relate psychophysical problems to physiology there is confusion about the level at which a problem arises -- is it related, for instance, mainly to the physical mechanisms of vision (like the after-images such as the one you see after staring at a lightbulb) or mainly to the computational theory of vision (like the ambiguity of the Necker cube?). More disturbingly, although the top level is the most neglected, it is also the most important. This is because the nature of the computations that underly perception depend more upon the computational problems that have to be solved than upon the particular hardware in which their solutions are implemented. To phrase the matter another way, an algorithm is likely to be understood more readily by understanding the nature of the problem that it deals with than by examining the mechanism (and the hardware) by which it is embodied. There is, after all, an analog to all of this in physics, where a thermodynamical approach represented, at least historically, the first stage in the study of matter: it succeeded in producing a theory of gross properties such as temperature. A description in terms of mechanisms or elementary components -- in this case atoms and molecules -- appeared some decades afterwards.

Our main point, therefore, is that the topmost of our four levels, that at which the necessary structure of computation is defined, is a crucial but neglected one. Its study is separate from the study of particular algorithms, mechanisms, or hardware, and the techniques needed to pursue it are new. In the rest of this article, we summarize some examples of vision theories at the uppermost level.

Conventional Approaches

The problems of visual perception have attracted the curiosity of scientists for many centuries. Important early contributions were made by Newton (1704), who laid the foundations for modern work on color vision, and Helmholtz (1910), whose treatise on physiological optics maintains its interest even today. Early in this century, Wertheimer (1923) noticed the apparent motion not of individual dots but instead of wholes, or "fields," in images presented sequentially, as if in a movie. In much the same way do we perceive the migration across the sky of a flock of geese, the flock somehow constituting a single entity, and not individual birds. This observation started the Gestalt school of psychology, which was concerned with describing the qualities of wholes, including solidarity and distinctness, and trying to formulate the laws that governed their creation. The attempt failed for various reasons, and the Gestalt school dissolved into the fog of subjectivism. With the death of the school, many of its early and genuine insights were unfortunately lost to the mainstream of experimental psychology.

The next developments of importance were recent and technical. The advent of electrophysiology in the 1940's and '50's made single cell recording possible, and with Kuffler's (1953) study of retinal ganglion cells -- the neurons of the eye that give rise to the optic nerve -- a new approach to the problem was born. Its most renowned practitioners are Hubel and Wiesel (1962, 1968), who since 1959 have conducted an influential series of investigations on single cell responses at various points along the visual pathway in the cat and the monkey. Students of the psychology of perception were also affected by a technological advance, the advent of the digital computer. Most notably, it allowed Bela Julesz in 1959 to devise random-dot stereograms (see Julesz 1971), which are image pairs constructed of dot patterns that appear random when viewed monocularly, but which fuse when viewed one through each eye to give a percept of shapes and surfaces with a clear three-dimensional structure. An example is shown in figure 1. Here the image for the left eye is a matrix of black and white squares generated at random by a computer program. The image for the right is made by copying the left image and then shifting a square-shaped region at its center slightly to the left, providing a new random pattern to fill in the gap that the shift must create. If each of the eyes sees only one matrix, as if they were both in the same physical place, the result is the sensation of a square floating in space. Plainly such percepts are caused solely by the stereo disparity between matching elements in the images presented to each eye. More recently, considerable interest has been attracted by a rather different approach. In 1971, Shepard and Metzler made line drawings of simple objects that differed from one another either by a three-dimensional rotation, or by a rotation plus a reflection (see figure 2). They asked how long it took to decide whether two depicted objects differed by a rotation and a reflection, or merely a rotation. They found that the time taken

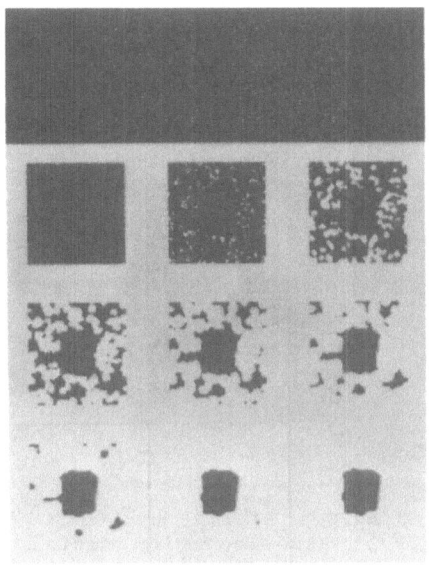

Figure 1: A random-dot stereogram (the top two images), and its decoding by Marr and Poggio's (1976) cooperative algorithm. The initial state contains all possible matches within a given disparity range, and the algorithm embodies the constraints of uniqueness and continuity to eliminate false targets. Shades of grey are used to signify matches at different disparities. The figure shows the initial state, and the states after 1, 2, 3, 4, 5, 6, 8 and 14 iterations. The algorithm progressively reveals a nested set of tiers. This algorithm is not the one used by the human visual system.

depended on the 3-D angle of rotation necessary to bring the two objects into correspondence. Indeed, it varied linearly with this angle. One is led thereby to the notion that a mental rotation of sorts is actually being performed: that a mental description of the first shape in a pair is being adjusted incrementally in orientation until it matches the second, such adjustment requiring greater time when greater angles are involved.

Interesting and important though these findings are, one must sometimes be allowed the luxury of pausing to reflect upon the overall trends that they represent, in order to take stock of the kind of knowledge that is accessible through these techniques. For we repeat:

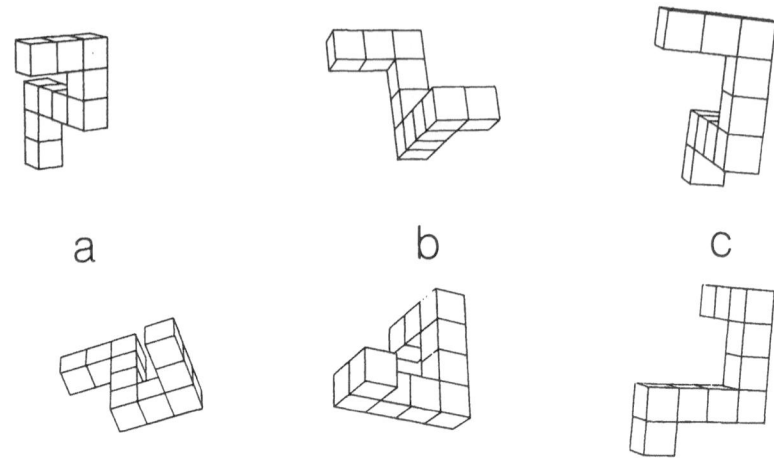

Figure 2: Some drawings similar to those used in Shepard and Metzler's (1971) experiments on mental rotation. Those shown in (a) and (b) are identical and the relative angle between the two is 80 degrees. Those in (c) are not identical, and no rotation will bring them into congruence.

perhaps the most striking feature of neurophysiology and psychophysics at present is that they _describe_ the behavior of cells or of subjects but do not _explain_ it. What are the visual areas of the cerebral cortex actually doing? What are the problems in doing it that need explaining, and at what level of description should such explanations be sought?

A Computational Approach to Vision

In trying to come to grips with these problems, our group at the M.I.T. Artificial Intelligence Laboratory has adopted a point of view that regards visual perception as a problem primarily in information processing. The problem commences with a large, gray-level intensity array, which suffices to approximate an image such as the world might cast upon the retinas of the eyes, and it culminates in a _description_ that depends on that array, and on the purpose that the viewer brings to it. Our particular concern in this article will be with the derivation of a description well-suited for the recognition of three-dimensional shapes.

The Primal Sketch. It is a commonplace that a scene and a drawing of the scene appear very similar, despite the completely different gray-level images to which they give rise. This suggests that the artist's symbols correspond in some way to natural symbols that are computed out of the image during the normal course of its interpretation. Our theory therefore asserts that the first operation on an image is to transform it into a primitive but rich description of the way its intensities change over the visual field, as opposed to a description of its particular intensity values in and of themselves. This yields a description of markedly reduced size that still captures the important aspects required for image analysis. We call it a primal sketch (Marr 1976). Consider, for example, an intensity array of 1,000 by 1,000, or a million points in all. Even if the possible intensity at any one point were merely black or white -- two different brightnesses -- the number of all possible arrays would still be $2^{1,000,000}$. In a real image, however, there tend to be continuities of intensity -- areas where brightness varies uniformly -- and this tends to eliminate possibilities in which the black and white oscillate wildly. It also tends to simplify the array. Typically, therefore, a primal sketch need not include a set of values for every point in an image. As stored in a computer, it will instead constitute an array with numbers representing the directions, magnitudes, and spatial extents of intensity changes assigned to certain specific points in an image -- points that tend to be places of locally high or low intensity. The positions of these points, particularly their arrangement amongst their immediate neighbors -- that is to say, the local geometry of the image -- must also be made explicit in the primal sketch, as it would otherwise be lost. (It was implicit, of course, in the 1,000-by-1,000 array, but we are no longer retaining data for each of those million places.) One way to do this is to specify "virtual lines" -- directions and distances -- between neighboring points of interest in the sketch.

The process of computing the primal sketch involves several steps. The first is the derivation of the raw primal sketch (see Marr and Hildreth 1979), which involves detecting and representing the intensity changes in the image. First, the image is filtered through a set of medium bandpass second differential operators $\nabla^2 G$, (where ∇^2 is the Laplacian and G is a Gaussian distribution), and the zero-

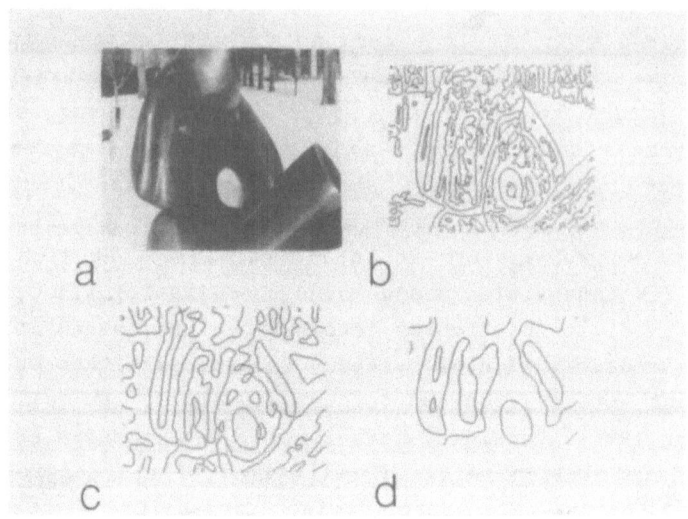

Figure 3: The image (a), which is 320 by 320 pixels, has been convolved with ^2G, a centre-surround operator with central excitatory region of width 2 = 6, 12 and 24 pixels. These filters span approximately the range of filters that operate in the human fovea. (b), (c), and (d) show the zero-crossings of the filtered images. These are the precursors of the raw primal sketch (from Marr and Hildreth 1979 figure 6).

crossings in the filtered images are found (see figure 3). This representation of the intensity changes is probably complete (Marr, Poggio and Ullman, 1979).

Although in general there is no reason why the zero-crossings found by the different channels should be related, in practice they will be. The reason is that most intensity changes in an image arise from physical phenomena that are spatially localized. This constraint allowed Marr and Hildreth to formulate the <u>spatial coincidence assumption</u> which states: If a zero-crossing is present in a set of independent ^2G channels over a contiguous range of sizes, and it has the same position and orientation in each channel, then the set of such zero-crossings may be taken to indicate the presence of an intensity change in the image that is due to a single physical

Recently, Marr and Ullman (1979) have extended the work of Marr and Hildreth to include the detection and use of directional selectivity. They have proposed specific roles for the X- and Y-channels found originally by Enroth-Cugell and Robson (1966), and in an explicit model for one class of cortical simple cell, they showed how to combine X and Y information to form a directionally selective unit.

Modules of Early Visual Processing The primal sketch of an image is typically a large and unwieldy collection of data, even despite its simplification relative to a gray-level array; for this is the unavoidable consequence of the irregularity and complexity of natural images. The next computational problem is thus its decoding. Now the traditional approach to machine vision assumes that the essence of such a decoding is a process called segmentation, whose purpose is to divide a primal sketch, or more generally an image, into regions that are meaningful, perhaps as physical objects. Tenenbaum and Barrow (1976), for example, applied knowledge about several different types of scene to the segmentation of images of landscapes, an office, a room, and a compressor. Freuder (1974) used a similar approach to identify a hammer in a simple scene. Upon finding a blob, his computer program would tentatively label it as the head of a hammer, and begin a search for confirmation in the form of an appended shaft. If this approach were correct, it would mean that a central problem for vision is arranging for the right piece of specialized knowledge to be made available at the appropriate time in the segmentation of an image. Freuder's work, for example, was almost entirely devoted to the design of a system that made this possible. But despite considerable efforts over a long period, the theory and practice of segmentation remain rather primitive, and here again we believe that the main reason lies in the failure to formulate precisely the goals of this stage of the processing -- a failure, in other words, to work at the topmost level of visual theory. What, for example, is an object? Is a head an object? Is it still an object if it is attached to a body? What about a man on horseback?

Marr (1976) argued that the early stages of visual information processing ought instead to squeeze the last possible ounce of information from an image before taking recourse to the descending influence of "high-level" knowledge about objects in the world. Let us turn, then, to a brief examination of the physics of the situation. As we noted earlier, the visual process begins with arrays of intensities projected upon the retinas of the eyes. The principal factors that

phenomenon (a change in reflectance, illumination , depth or surface orientation).

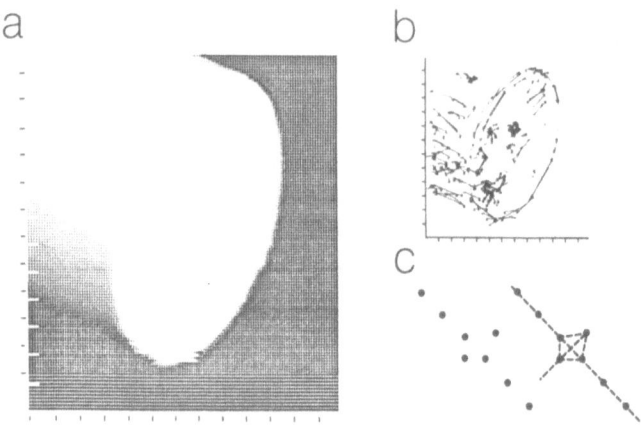

Figure 4: The primal sketch makes explicit information held in an intensity array (a). There are two kinds of information: (b) one concerns changes in intensity, represented by oriented edge, bar and blob primitives,together with associated parameters that measure the contrast and spatial extent of the intensity change; and the other (c) is the localgeometry of significant places in the image. Such places are marked by place-tokens, which can be defined in a variety of ways, and the geometric relations between them are represented by virtual lines (Marr 1976 figures 7 and 12a).

This assumption allows one to combine the zero-crossings from different channels into edge-segment descriptors, bars and blobs (see figure 4b), which constitute the raw primal sketch. To obtain the full primal sketch, these primitive elements are grouped, perhaps hierarchically, into units called place-tokens, which associate properties like length, width, brightness and so forth with positions in the image (Marr 1976). Virtual lines may then be used to represent the local geometry of these place tokens (see figure 4c and Stevens, 1978).

determine these intensities are (1) the illuminant, (2) the surface reflectance properties of the objects viewed, (3) the shapes of the visible surfaces of these objects, and (4) the vantage point of the viewer. Thus if the analysis of the input intensity arrays is to operate autonomously, at least in its early stages, it can only be expected to extract information about these four factors. In short, early visual processing must be limited to the recovery of localized physical properties of the visible <u>surfaces</u> of a viewed object -- particularly local surface dispositions (orientation and depth) and surface material properties (color, texture, shininess, and so on). More abstract matters such as a description of overall three-dimensional shape must come after this more basic analysis is complete.

An example of early processing is stereopsis. Imagine that images of a scene are available from two nearby points at the same horizontal level -- the analog of the images that play upon the retinas of your left and right eyes. The images are somewhat different, of course, in consequence of the slight difference in vantage. Imagine further that a particular location on a surface in the scene is chosen from one image; that the corresponding location is identified in the other image; and that the relative positions of the two versions of that location are measured. This information will suffice for the calculation of depth -- the distance of that location from the viewer. Notice that methods based on gray-level correlation between the pair of images fail to be suitable because a mere gray-level measurement does not reliably define a point on a physical surface. To put the matter plainly, numerous points in a surface might fortuitously be the same shade of gray, and differences in the vantage points of the observer's eyes could change the shade as well. The matching must evidently be based instead on objective markings that lie upon the surface, and so one has to use changes in reflectance. One way of doing this is to obtain a primitive description of the intensity changes that exist in each image (such as a primal sketch), and then to match these descriptions. After all, the line segments, edge segments, blobs, and edge termination points included in such a description correspond quite closely to boundaries and reflectance changes on physical surfaces. The stereo problem -- the determination of depth given a stereo pair of images -- may thus be reduced to that of matching two primitive descriptions, one from each eye; and to help in this task there are physical constraints that translate into two rules for how the left and right descriptions are combined:

Uniqueness. Each item from each image may be assigned at most one disparity value -- that is to say, a unique position relative to its counterpart in the stereo pair. This condition rests on the premise that the items to be matched have a physical existence, and can be in only one place at a time.

Continuity. Disparity varies smoothly almost everywhere. This condition is a consequence of the cohesiveness of matter, and it states that only a relatively small fraction of the area of an image is composed of discontinuities in depth.

In the case of random-dot stereograms, the computational problem is rather well-defined, essentially because of Julesz's demonstration that random-dot stereograms, containing no monocular information, still yield stereopsis. In 1976, Marr and Poggio developed a method for computing local disparities in a pair of random-dot stereograms by an iterative, parallel procedure known technically as a cooperative algorithm (see figure 1, and Marr, Poggio and Palm 1977). This sort of algorithm has the property that it can be defined completely in terms of simple local interactions because at each of its iterations, each point is affected only by a calculation performed on its immediate neighborhood. Yet all points are so affected during each successive iteration, so the transformations take on a complex global nature. Subsequent comparison of the algorithm's performance with psychophysical data showed that it did not hold up well as a model for human stereopsis. To be sure, it performed better than people do on the standard stereograms like that shown in figure 1; but it did not explain people's ability to see stereograms in which one of the two images is defocused slightly or enlarged slightly relative to the other. These observations led Marr and Poggio (1979) to devise another algorithm, this one based on the human use of spatial-frequency-tuned channels and vergence eye movements. This algorithm is consistent with all of the currently known psychophysical data.

A second example of early visual processing concerns the derivation of structure from motion. It has long been known that as an object moves relative to the viewer, the way its appearance changes provides information that we can use to determine its shape (Wallach and O'Connell 1953). The motion analog of a random-dot stereogram is illustrated in figure 5, and as expected, humans can easily perceive shape from a succession of frames, each of which on its own is merely a set of random-dots. In various papers and a forthcoming book on the

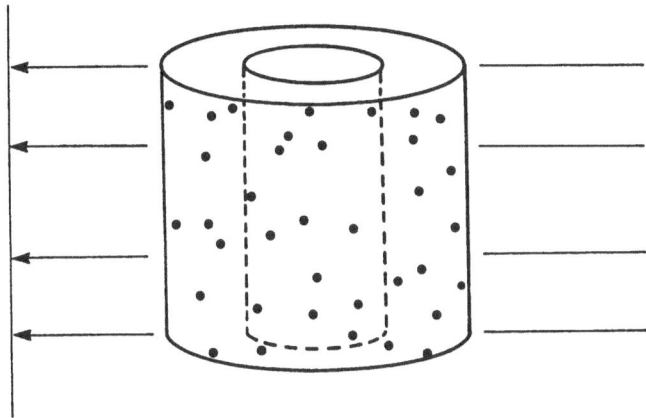

Figure 5: The motion analogue of the random-dot stereogram. Two transparent,concentric cylinders are rotated in opposite directions. Each has dots scattered on its surface. A movie camera photographs the scene from the side, and each frame contains only a pattern of random dots. When a human watches the movie, however, he immediately perceives the two counter-rotating cylinders (from Ullman 1979).

subject, Ullman (1979a and b) decomposed the problem into two parts: matching the elements that occur in consecutive images; and deriving shape information from measurements of their changes in position. Ullman then showed that these problems can be solved mathematically. His basic idea is that in general, nothing can be inferred about the shape of an object given only a set of sequential views of it; for some extra assumptions have to be made. Accordingly, he formulates an assumption of rigidity, which states that if a set of moving points has a _unique_ interpretation as a rigid body in motion, that interpretation is correct. (The assumption is based on a theorem which he proves, stating that three distinct views of four noncoplanar points on a rigid body are sufficient to determine uniquely their three-dimensional arrangement in space.) From this he derives a method for computing structure from motion. The method gives results that are quantitatively superior to the ability of humans to determine shape from motion, and which fail in qualitatively similar circumstances. Ullman has also devised a set of simple algorithms by which the method may be implemented.

The 2 1/2-Dimensional Sketch. Both of the techniques of image analysis discussed in the preceding paragraphs provide information about the relative distances to various .places in an image. In the case of stereopsis, it is the matching of points in a stereo pair that leads to such information. In the case of structure from motion, it is the matching of points in successive images. More generally, however, we know that vision provides several sources of information about shapes in the visual world. The most direct, perhaps, are the aforementioned stereo and motion, but texture gradients in a single image are nearly as effective. Furthermore, the theatrical techniques of facial make-up reveal the sensitivity of perceived shapes to shading (see Horn 1975), and color sometimes suggests the manner in which a surface reflects light. It often happens that some parts of a scene are open to inspection by some of these techniques, and other parts to inspection by others. Yet different as the techniques are, they all have two important characteristics in common: they rely on information from the image rather than a priori knowledge about the shapes of the viewed objects; and the information they specify concerns the depth or surface orientation at arbitrary points in an image, rather than the depth or orientation associated with particular objects (see Table 1).

Information source	Natural parameter
Stereo	Disparity, hence especially δr and Δr
Motion	r, hence δr, Δr
Shading	$\underset{\sim}{s}$
Texture gradients	$\underset{\sim}{s}$
Perspective cues	$\underset{\sim}{s}$
Occlusion	Δr
Contour	$\underset{\sim}{s}$

Table 1: The form in which various early visual processes deliver information about the changes in a scene.

$$r = \text{depth}$$
$$\delta r = \text{small, local changes in depth}$$
$$\Delta r = \text{large changes in depth}$$
$$\underset{\sim}{s} = \text{local surface orientation}$$

In order to make the most efficient use of different and often complementary channels of information deriving from stereopsis, from motion, from contours, from texture, from color, from shading, they need to be combined in some way. The computational question that now arises is thus how best to do this, and the natural answer is to seek some representation of the visual scene that makes explicit just the information these processes can deliver. We seek, in other words, a representation of surfaces in an image that makes explicit their shapes and orientations, much as the Arabic representation of a number makes explicit its composition by powers of ten. It might be contrasted with the representation of a surface as a mathematical expression, in which the orientation is only implicit, and not at all apparent. We call such a representation the 2 1/2-dimensional sketch (Marr and Nishihara 1978; Marr 1978), and in the particular candidate for it shown in figure 6, surface orientation is represented by covering an image with needles. The length of each needle defines the dip of the surface at that point, so that zero length corresponds to a surface that is perpendicular to the vector from the viewer to the point, and increasing lengths denote surfaces that tilt increasingly away from the viewer. The orientation of each needle defines the local direction of dip.

Our argument is that the 2 1/2-D sketch is useful because it makes explicit information about the image in a form that is closely matched to what image analysis can deliver. To put it another way, we can formulate the goals of this stage of visual processing as being primarily the construction of this representation, discovering, for example, what are the surface orientations in a scene, which of the contours in the primal sketch correspond to surface discontinuities and should therefore be represented in the 2 1/2-D sketch, and which contours are missing in the primal sketch and need to be inserted into the 2 1/2-D sketch in order to bring it into a state that is consistent with the nature of three-dimensional space. This formulation avoids the difficulties associated with the terms "region" and "object" -- the difficulties inherent in the image segmentation approach; for the gray level intensity array, the primal sketch, the various modules of early visual processing, and finally the 2 1/2-dimensional sketch itself deal only with discovering the properties of surfaces in an image. One is pleased about that, for we know of ourselves as perceivers that surface orientation can be associated with unfamiliar shapes, so its representation probably precedes the decomposition of the scene into objects. One is thus free to ask

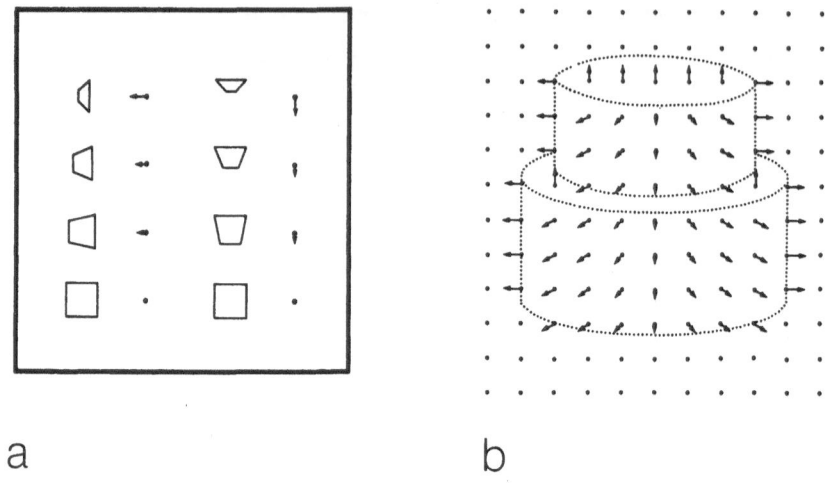

a   b

Figure 6: Illustration of the 2 1/2-dimensional sketch. In (a) the perspective views of small squares placed at various orientations to the viewer are shown. The dots with arrows show a way of representing the orientations of such surfaces symbolically. In (b), this representation is used to show the surface orientations of two cylindrical surfaces in front of a background orthogonal to the viewer. The full 2 1/2-dimensional sketch would include rough distances to the surfaces as well as their orientations, contours where surface orientation changes sharply, and contours where depth is discontinuous (subjective contours). A considerable amount of computation is required to maintain these quantities in states that are consistent with one another and with the structure of the outside world (see Marr 1978 section 3). (From Marr and Nishihara 1978 figure 2).

precise questions about the computational structure of the 2 1/2-D sketch and of processes to create and maintain it. We are currently much occupied with these matters.

Later Processing Problems

The final components of our visual processing theory concern the application of visually derived surface information for the representation of three-dimensional shapes in a way that is suitable specifically for recognition (Marr and Nishihara 1978). By this we mean the ability to recognize a shape as being the same as a shape seen earlier, and this in essence depends on being able to describe shapes consistently each time they are seen, whatever the circumstances of their positions relative to the viewer. The problem with local surface representations such as the 2 1/2-D sketch is that the description depends as much on the viewpoint of the observer as it does on the structure of the shape. In order to factor out a description of a shape that depends on its structure alone, the representation must be based on readily identifiable geometric features of the overall shape, and the dispositions of these features must be specified relative to the shape in itself. In brief, the coordinate system must be "object-centered," not "viewer-centered." One aspect of this deals with the nature of the representation scheme that is to be used, and another with how to obtain it from the 2 1/2-D sketch. We begin by discussing the first, and will then move on to the second.

The 3-D Model Representation. The most basic geometric properties of the volume occupied by a shape are (1) its average location (or center of mass); (2) its overall size, as exemplified, for example, by its mean diameter or volume; and (3) its principal axis of elongation or symmetry, if one exists. A description based on these qualities would certainly be inadequate for an application such as shape recognition; after all, one can tell little about the three-dimensional structure of a shape given only its position, size, and orientation. But if a shape itself has a natural decomposition into components that can be so described, this volumetric scheme is an effective means for describing the relative spatial arrangement of those components. The illustration of figure 7 shows a familiar version of this type of description, the stick figure (see Blum 1973). The recognizability of the animal shapes depicted in the illustration is surprising considering the simplicity of representation used to describe them.

The reason such a description works so well lies, we think, in (1) the volumetric (as opposed to surface-based) definition of the primitive elements -- the sticks -- used by the representation; (2) the relatively small number of elements used; and (3) the relation of elements to each other rather than to the viewer. In short, this type of shape representation is volumetric, modular, and can be based on object-centered coordinates. The figure 8 illustrates the scheme of representation that was developed from these ideas. Here the

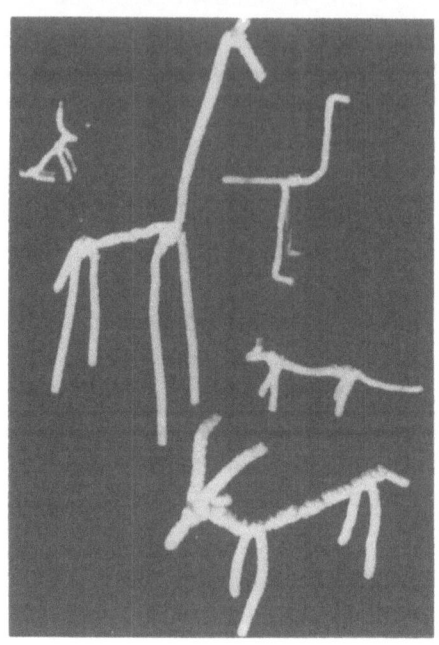

Figure 7: The portrayal of animals by a small number of pipecleaners serves to show that the representation of a three-dimensional shape need not make explicit its surface in order to describe it so well that it can easily be recognised. The success of the representation is due, one suspects, in large measure to the correspondence between the pipecleaners and the axes of the volumes they stand for. (From Marr and Nishihara 1978 figure 13).

description of a shape is composed of a hierarchy of stick-figure specifications we call 3-D models. In the simplest, a single axis element is used to specify the location, size, and orientation of the entire shape; the human body displayed in the illustration will serve as an instance. This element is also used to define a coordinate system that will specify the dispositions of subsidiary axes, each of these specifying in turn a coordinate system for 3-D models of "arm," "hand," and so on. This hierarchical structure makes it possible to treat any component of a shape as a shape in itself. It also provides flexibility in the detail of a description.

Shapes Admitting 3-D Model Descriptions. If the scheme for a given shape is to be uniquely defined and stable over unimportant variations such as viewpoint -- if, in a word it is to be canonical -- its

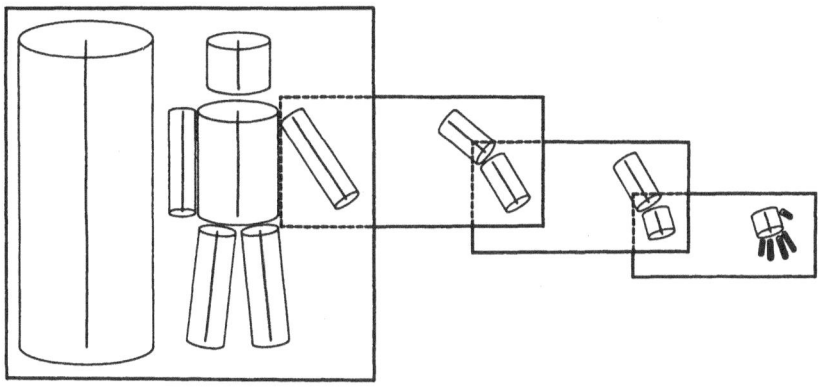

Figure 8: The arrangement of 3-D models into the
representation of a human shape. First the overall form --
the "body" -- is given an axis. This yields an object-
centred coordinate system which can then be used to specify
the arrangement of the "arms," "legs," "torso" and "head."
The position of each of these is specified by an axis of its
own, which in turn serves to define a coordinate system for
specifying the arrangement of further subsidiary parts. This
gives us a hierarchy of 3-D models, shown here extending
downwards as far as the fingers. The shapes in the figure
are drawn as if they were cylindrical, but that is purely
for illustrative convenience. (From Marr and Nishihara 1978
figure 3).

definition must take advantage of any salient geometrical
characteristics that the shape inherently possesses. If a shape has
natural axes, then those should be used. The coordinate system for a
sausage should take advantage of its major axis, and for a face, of
its axis of symmetry.

Highly symmetrical objects, like a sphere, a square, or a circular
disc, will inevitably lead to ambiguities in the choice of coordinate
systems. For a shape as regular as a sphere this poses no great
problem, because its description in all reasonable systems is the
same. One can even allow óther factors, like the direction of motion

or spin, to influence the choice of coordinate frame. For other
shapes, the existence of more than one possible choice probably means
that one has to represent the object in several ways, but this is
acceptable provided that their number is small. For example, there are
four possible axes on which one might wish to base the coordinate
system for representing a door, namely the midlines along its length,
its width, and its thickness, and also the axis of its hinges. (This
last would be especially useful to represent how the door opens.) For
a typewriter, there are two reasonable choices, an axis parallel to
its width, because that is usually its largest dimension, and the axis
about which a typewriter is roughly symmetrical.

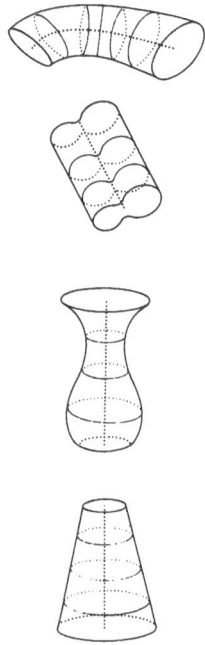

Figure 9: The definition of a generalized cone. It is the
surface created by moving a cross-section along a given
straight axis. The cross-section may vary smoothly in size,
but its shape remains constant. Several examples are shown
here. In each, the cross-section is shown at several
positions along the trajectory that spins out the
construction.

In general, if an axis can be distinguished in a shape, it can be
used as the basis for a local coordinate system. One approach to the
problem of defining object-centered coordinates is therefore to

examine the class of shapes having an axis as an integral part of their structure. Consider, accordingly, the class of so-called generalized cones, each of these being the surface swept out by moving a cross-section of constant shape but smoothly varying size along an axis, as shown in figure 9. Binford (1971) has drawn attention to this class of constructions, suggesting that it might provide a convenient way of describing three-dimensional surfaces for the purposes of computer vision (see also Agin, 1972; Nevatia, 1974). We regard it as an important class not because the shapes themselves are easily describable, but because the presence of an axis allows one to define a canonical local coordinate system. Fortunately, many objects, especially those whose shape was achieved by growth, are described quite naturally in terms of one or more generalized cones. The animal shapes of figure 7 provide some examples; the individual sticks are simply the axes of generalized cones that approximate the shapes of parts of these creatures. Many artifacts can also be described in this way -- say a car (a small box sitting atop a longer one) or a building (a box with a vertical axis).

It is important to remember, however, that there exist surfaces that cannot conveniently be approximated by generalized cones, for example a cake that has been transected at some arbitrary plane, or the surface formed by a crumpled newspaper. Cases like the cake could be dealt with by introducing a suitable surface primitive for describing the plane of the cut, in much the same way as an axis in the 3-D model representation is a primitive that describes a volumetric element. But the crumpled newspaper poses apparently intractable problems.

Finding the Natural Coordinate System. Even if a shape possesses a canonical coordinate frame, one still is faced with the problem of finding it from an image. Our own interest in this problem grew from the question of how to interpret the outlines of objects as seen in a two-dimensional image (Marr 1977b), and our starting point was the observation that when one looks at the silhouettes in Picasso's "Rites of Spring" (reproduced here in figure 10), one perceives them in terms of very particular three-dimensional shapes, some familiar, some less so. This is quite remarkable, because the silhouettes could in theory have been generated by an infinite variety of three-dimensional shapes which, from other viewpoints, would have no discernible similarities to the shapes we perceive. One can perhaps attribute part of the phenomenon to a familiarity with the depicted shapes, but not all of

Figure 10: "Rites of Spring" by Pablo Picasso. We
immediately interpret such silhouettes in terms of
particular three-dimensional surfaces -- this despite the
paucity of information in the image itself. In order to do
this we plainly must invoke certain a priori assumptions and
constraints about the nature of the shapes.

it, because one can use the medium of a silhouette to convey a new
shape, and because even with considerable effort it is difficult to
imagine the more bizarre three-dimensional surfaces that could have
given rise to the same silhouettes. The paradox, then, is that the
bounding contours in Picasso's "Rites" apparently tell us more than
they should about the shape of the figures. For example, neighboring
points on such a contour could in general arise from widely separated
points on the original surface, but our perceptual interpretation
usually ignores this possibility.

The first observation to be made is that the contours that bound
these silhouettes are contours of surface discontinuity, which are
precisely the contours with which the 2 1/2-D sketch is concerned.
Secondly, because we can interpret the silhouettes as three-
dimensional shapes, then implicit in the way we interpret them must

lie some a priori assumptions that allow us to infer a shape from an
outline. If a surface violates these assumptions, our analysis will be
wrong, in the sense that the shape we assign to the contours will
differ from the shape that actually caused them. An everyday example
is the shadowgraph, where the appropriate arrangement of one's hands
can, to the surprise and delight of a child, produce the shadow of a
duck or a rabbit.

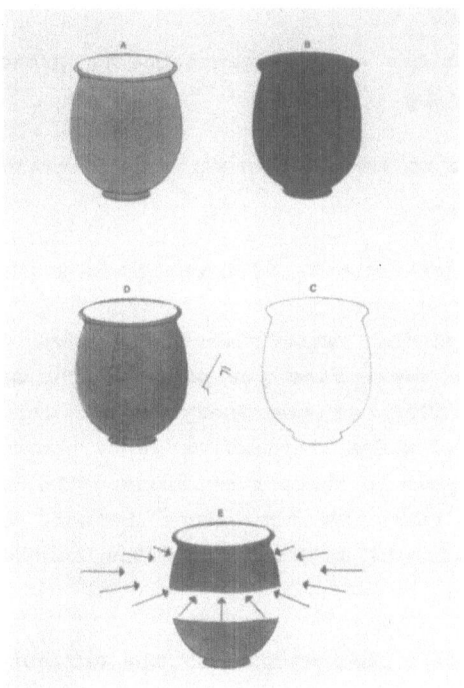

Figure 11: Four structures of importance in studying the a
priori conditions mentioned in figure 10. (a) shows a three-
dimensional surface Σ. (b) shows its silhouette S_V as seen
from viewpoint V. (c) shows the contour C_V of S_V, and (d)
shows the set of points Γ_V on Σ that project onto the
contour. Finally, (e) illustrates schematically the meaning
of the phrase "all distant viewing directions that lie in a
plane."

What assumptions is it reasonable to suppose that we make? In
order to explain them, we need to define the four constructions that
appear in figure 11. These are (1) a three-dimensional surface Σ; (2)
its image or silhouette S_V as seen from a viewpoint V; (3) the
bounding contour C_V of S_V; and (4) the set of points on the surface
Σ that project onto the contour C_V. We shall call this last the
contour generator of C_V, and we shall denote it by Γ_V.

Observe that the contour C_V, like the contours in the work of Picasso, imparts very little information about the three-dimensional surface that caused it. Indeed, the only obvious feature available in the contour is the distinction between convex and concave places -- that is to say, the presence of inflection points. In order that these inflections be "reliable," one needs to make some assumptions about the way the contour was generated, and we choose the following restrictions (Marr 1977):

1. Each point on the contour generator Γ_V projects to a different point on the contour C_V.

2. Nearby points on the contour C_V arise from nearby points on the contour generator Γ_V.

3. The contour generator Γ_V lies wholly in a single plane.

The first and second restrictions say that each point on the contour of the image comes from one point on the surface (which is an assumption that facilitates the analysis but is not of fundamental importance), and that where the surface looks continuous in the image, it really is continuous in three dimensions. The third restriction is simply the demand that the difference between convex and concave contour segments reflects properties of the surface, rather than of the imaging process.

It turns out to be a theorem that if the surface is smooth (for our purposes, if it is twice differentiable with continuous second derivative) and if restrictions 1 through 3 hold for all distant viewing positions in any one plane (as illustrated in figure 11), then the viewed surface is a generalized cone. (The converse is also true: if the surface is a generalized cone, then conditions 1 through 3 will be found to be true).

This means that if the convexities and concavities of a bounding contour in an image are actual properties of a surface, then that surface is a generalized cone or is composed of several such cones. In brief, the theorem says that a natural link exists between generalized cones and the imaging process itself. The combination of these two must mean, we think, that generalized cones will play an intimate role in the development of vision theory.

Discussion

We have tried in this survey of visual information processing to
make two principal points. The first is methodological: namely that it
is important to be very clear about the nature of the understanding we
seek. The results we try to achieve should be precise ones, at the
level of what we call a computational theory. The critical act in
formulating computational theories turns out to be the discovery of
valid constraints on the way the world is structured -- constraints
that provide sufficient information to allow the processing to
succeed. Consider stereopsis, which presupposes continuity and
uniqueness in the world, or structure from visual motion, which
presupposes rigidity, or shape from contour, which presupposes the
three restrictions just discussed, or even edge detection, which
presupposes the assumption of spatial coincidence. The discovery of
constraints that are valid and universal leads to results about vision
that have the same quality of permanence as results in other branches
of science.

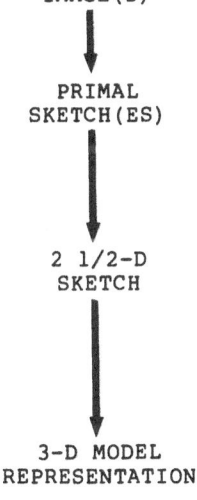

IMAGE(S)

PRIMAL
SKETCH(ES)

Describes the intensity changes present in
an image, labels distinguished locations
like termination points, and makes explicit
local two-dimensional geometrical relations.

2 1/2-D
SKETCH

Represents contours of surface discontinuity,
and depth and orientation of visible surface
elements, in a coordinate frame that is centered on the viewer.

3-D MODEL
REPRESENTATION

Shape description that includes volumetric
shape primitives of a variety of sizes,
whose positions are defined using an object-
centered coordinate system. This representation imposes considerable modular
organization on its descriptions.

Table 2: A framework for the derivation of shape
information from images.

The second point is that the critical issues for vision seem to us
to revolve around the nature of the representations and the nature of

the processes that create, maintain, and eventually interpret them. We have suggested an overall framework for visual information processing (summarized in table 2), that includes three categories of representation upon which the processing is to operate. The first encompasses representations of intensity variations and their local geometry in the input to the visual system. One among these, the primal sketch, is expressly intended to be an efficient description of these variations which captures just that information required by the image analysis to follow. The second category encompasses the representations of visible surfaces -- the descriptions, in other words, of the physical properties of the surfaces that caused the images in the first place. The nature of these representations -- the 2 1/2-dimensional sketch in particular -- is determined primarily by what information can be extracted by modules of image analysis such as stereopsis and structure from motion. Like the primal sketch of the previous category, the 2 1/2-dimensional sketch is intended to be a final or output representation: this is where the separate contributions from the various image-analysis modules can be combined into a unified description. The third category encompasses all representations which are subsequently constructed from information contained in the 2 1/2-D sketch. The designs of these tertiary representations are determined largely by the use to which they are to be put, as was the case for the 3-D model representation, to be used for shape recognition. If one had wanted instead, for example, to represent a shape simply for later reproduction, say by the milling of a block of metal, then the 2 1/2-D sketch would itself have been sufficient, as the milling process depends explicitly on information about local depth and orientation, such as that sketch can provide.

Finally, a remark of a rather different nature. As we have seen, some aspects of human early visual processing, like stereopsis, have apparently been understood well enough to implement them in machines (Marr and Poggio, 1979; Marr and Grimson, 1979). The computational power required by these early processes is prohibitive, and until recently the prospects for real-time implementation of human-like early vision were remote. It now appears, however, that the emerging VLSI and CCD technologies will be able to supply the necessary processing power. This could make the next two decades very interesting.

ACKNOWLEDGEMENTS: I thank Keith Nishihara and Michael Feiertag for help in assembling this survey. I thank the Royal Society for permission to reproduce figures 3, 4, 6, 7, and 8, and the M.I.T. press for figure 5. This work was conducted at the Artificial Intelligence Laboratory, a Massachusetts Institute of Technology research program supported in part by the advanced Research Projects Agency of the Department of Defense, and monitored by the Office of Naval Research under contract number N00014-75-C-0643, and in part by NSF contract number 77-07569-MCS.

References

Agin, G.J. 1972. Representation and description of curved objects. Stanford Artificial Intelligence Project, Memo AIM-173, Stanford University.

Binford, T.O. 1971. Visual perception by computer. Presented to the IEEE Conference on Systems and Control, Miami, December.

Blum, H. 1973. Biological shape and visual science, (part 1). J. theor. Biol., 38. 205-287.

Freuder, E.C. 1975. A computer vision system for visual recognition using active knowledge. M.I.T.A.I. Lab. Technical Report 345.

Helmholtz, H.L.F. von 1910. Treatis on physiological optics. Translated by J.P. Southall, 1925, N.Y. Dover Publications.

Horn, B.K.P. 1975. Obtaining shape from shading information. In The Psychology of Computer Vision, Ed. P.H. Winston. McGraw-Hall, New York, pp115-155.

Hubel, D.H. and Wiesel, T.N. 1962. Receptive fields and functional architecture of monkey striate cortex. J. Physiol.(Lond.) 195, 215-243.

Julesz, B. 1971. Foundations of Cyclopean Perception. Chicago: The University of Chicago Press.

Kuffler, S.W. 1953. Discharge patterns and functional organization of mammalian retina. J. Neurophysiol. 16, 37-68.

Marr, D. 1976. Early processing of visual information. Phil. Trans.Roy.Soc.B. 275, 483-524.

Marr, D. 1977a. Artificial Intelligence - a personal view. Artificial Intelligence 9, 37-48.

Marr, D. 1977b. Analysis of occluding contour. Proc.Roy.Soc.B. 197, 441-475.

Marr, D. 1978. Representing visual information. Lectures on mathematics in the life sciences, volume 10, Some Mathematical Questions in Biology, 101-180.

Marr, D., and Hildreth, E. (1979). Theory of edge detection. M.I.T.A.I. Memo #518.

Marr, D. and Nishihara, H.K. 1978. Representation and recognition of the spatial organization of three-dimensional shapes. Proc.Roy.Soc.B. 200, 269-294.

Marr, D. and Poggio, T. 1976. From understanding computation to understanding neural circuitry. Neurosciences Res. Prog. Bull. 15, 470-488.

Marr, D., Poggio, T. and Palm, G. 1977. Analysis of a cooperative stereo algorithm. Biol. Cybernetics 28, 223–239.

Marr, D. and Poggio, T. 1979. A theory of human stereo vision. Proc.Roy.Soc.Lond. (in the press).

Marr, D. Poggio, T. and Ullman, S.1979. Bandpass channels, zero-crossings and early visual information processing. J. opt.Soc.Am., (in the press).

Marr, D. and Ullman, S. 1979. Directional selectivity and its use in early visual processing. (In preparation).

Nevatia, R. 1974. Structured descriptions of complex curved objects for recognition and visual memory. Stanford Artificial Intelligence Project, Memo AIM-250, Stanford University.

Newton, I. 1704. Optics. London.

Shepard, R.N. and Metzler, J. 1971. Mental rotation of three-dimensional objects. Science. 171, 701-703.

Stevens, K.A. 1978. Computation of locally parallel structure. Biol.Cybernetics 29, 19-28.

Tenenbaum, J.M. and Barrow, H.G. 1976. Experiments in interpretation-guided segmentation. Stanford Research Institute Technical Note 123.

Ullman, S. 1979a. The interpretation of structure from motion. Proc.Roy.Soc.Lond. (in the press).

Ullman, S. 1979b. The interpretation of visual motion. M.I.T. press, March.

Trigger Features or Fourier Analysis in Early Vision: A New Point of View

T. Poggio

Max-Planck-Institut für biologische Kybernetik
74 Tubingen 1, Spemannstrasse 38, Germany

with an appendix by H.K. Nishihara
M.I.T. Artificial Intelligence Laboratory
545 Technology Square, Cambridge MA 02139

Abstract: Under the appropriate conditions, zero-crossings of bandpass signal are very rich in information (Logan, 1977). We examine here the relevance of this result to the early stages of visual information processing, where zero-crossings in the output of independent spatial-frequency-tuned channels may contain sufficient information for most of the subsequent processing.

The pioneering experiments of Hubel and Wiesel (1962) and of Campbell and Robson (1968) introduced two rather distinct notions of the function of early information processing in higher visual systems. Hubel and Wiesel's description of simple cells with bar- or edge-shaped receptive fields led to a view of the cortex as containing a population of feature detectors tuned to edges and bars of various widths and orientations. The extreme form of this view is represented by the notion of "grandmother" of even "pontifical" cells at the top of an hierarchy in which "bar detectors" are, in vision, one of the first stages. Campbell and Robson's experiments, showing that visual information is processed in parallel by a number of independent spatial-frequency-tuned channels, suggested a rather different view, which in its extreme form describes the visual cortex as a kind of spatial Fourier analyzer.

Neither of these views bears very close examination. The main points against a Fourier interpretation are: (1) The bandwidth of the channels is not very narrow. It is impossible to represent "Fourier coefficients" by means of cells with spatially localized receptive fields. (2) As Campbell and Robson found, early visual information processing is not linear (e.g. probability summation [Wilson and Geize, 1977] and failure of superposition). (3) No convincing demonstration has yet been made showing that phase information is coded. The notion that a simple cell with a bar-shaped receptive field acts as a bar-detector is simplistic. Such cells respond to many other stimuli as well. To build a truly reliable and selective bar-detector would solve a fair amount of the problems of early visual information processing.

Perhaps the most unsatisfactory aspect of the feature detector or Fourier interpretations of these early experiments is that they give little clue to the underlying information processing taking place, and the goals of early analysis of the image. This motivated a new approach to vision, which enquired directly about the information processing problems inherent in the task itself (Marr, 1976). According to this scheme, the purpose of early visual processing is to construct a primitive but rich description of the image that is ultimately to be used to determine the orientation and relative depth in space of the visible surfaces, relative to the viewer. This first primitive description of the image was called the primal sketch (Marr, 1976) and it concentrates on representing changes in intensity and local geometry, on the grounds that these correspond with high probability to physically important items like the markings on a surface, and their geometrical arrangement there. In most approaches to computer vision, an important preliminary computation is in fact the localization of discontinuities in image intensity. This can be achieved by finding peaks in the first directional derivative of intensity, or equivalently, zero-crossings in the second directional derivative. The latter quantity may be obtained by convolving the image with a bar-shaped mask, which approximates the second directional derivative at its particular scale. By using a range of mask sizes, one can begin to deal with the wide range of scales over which changes take place in a natural image.

These ideas begin to account, on purely information processing grounds, for the presence of spatial-frequency-tuned channels in early human vision. Recent work by Wilson and Gieze (1977) shows that such channels can be realized by linear units with bar-shaped receptive fields, reminiscent of the simple cells that Hubel and Wiesel (1962) have described. Marr and Poggio's (1977, 1979) recent theory of human stereopsis is, for example, conceived within this framework, and assumes that the elements that are matched between the two images are equivalent to the zero-crossings in bar-mask outputs. The object of this chapter is to point out that very recent advances in information

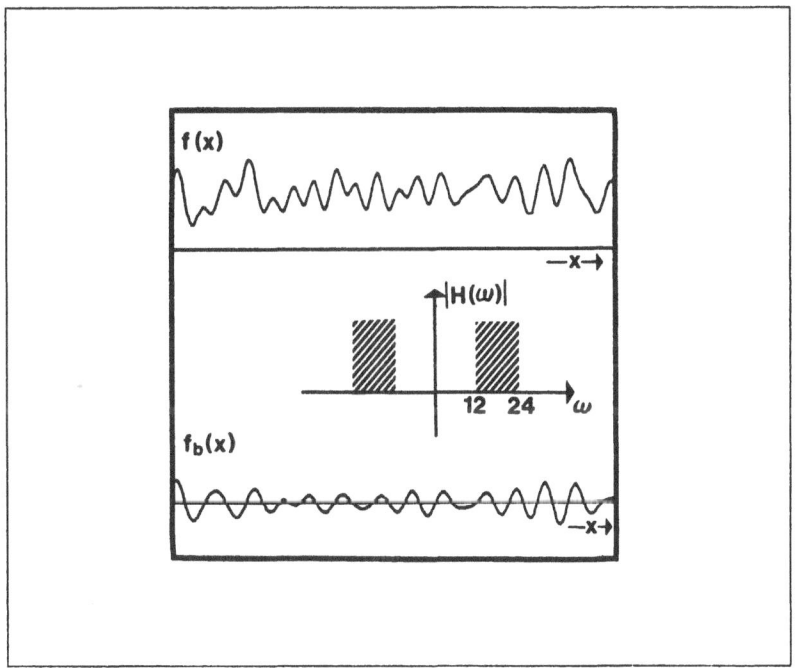

Figure 1. The meaning of Logan's theorem. (a) shows a stochastic bandlimited Gaussian signal $f(x)$, and (c) exhibits the result $f_b(x)$ of filtering (a) through an ideal one-octave bandpass filter. The modulus of its transfer function is shown in (b). Since (c) has a bandwidth of one octave, and it has only simple real zeros and no complex zeros in common with its Hilbert transform, Logan's theorem tells us that (c) is determined, up to a multiplicative constant, by its zero-crossings alone. The aspect of Logan's result that is important for this article is that under the right conditions, zero-crossings alone are very rich in information.

theory provide fascinating additional theoretical support for this framework.

The advance in question is a theorem proved by Logan (Logan, 1977) who showed that if a one-dimensional analytic function is (a) bandpass, with enough real zeros in proportion to its bandwidth and (b) has no free zeros, (i.e. complex zeros in common with its Hilbert transform) and no multiple real zeros, then the function is completely determined (up to an overall multiplicative constant) by its (real) zero-crossings (see figure 1). Condition (a) is critical, but condition (b) can for practical purposes be ignored, since it is almost always satisfied except by pathological signals. Condition (a) is always ensured if the bandwidth of the signal is less than one octave. (For an illustration of Logan's results in the special case of trigonometric polynomials, see Appendix 1).

If one translates this result into the context of early visual processing, its meaning is this. We have already seen that the basic idea of using zero-crossings in bar-mask convolutions from which to generate a primitive description of the image has a strong physical motivation. Logan's result tells us that, if the bar-mask operators are band-pass with a narrow enough bandwidth, then the zero-crossings alone are so rich in information that they determine essentially completely the convolution values (taken along a scan line perpendicular to the mask orientation). This method of recovering scan lines is the most obvious way of applying Logan's inherently one-dimensional result to two dimensional images. It requires that each one-dimensional function along a scan line be bandpass, for instance, with a one-octave bandwidth. Note that this is not necessarily satisfied by filtering the image through a two-dimensional bandpass filter (like a ring in the (ω_x, ω_y) plane of width one octave). For instance, an image filtered through a (bandpass) bar-shaped mask is bandpass on each scan-line perpendicular to the mask's orientation; an image filtered through a (bandpass) circularly-symmetric mask is band-limited but not bandpass along any scan-line. This follows from the fact that the Fourier transform along (for instance) the x-axis of an image filtered through a bandpass "ring" is essentially the projection of the two-dimensional Fourier transform on ω_x, and is therefore not bandpass.

Within this framework of a rather direct application of Logan's theorem one is committed to orientation-dependent operators applied along scan lines across the image. Figure 2 illustrates how a two-dimensional image can be reconstructed from zero-crossings along scan lines. This method relies upon transforming the two-dimensional image into a set of one-dimensional functions to which Logan's theorem can then be applied. Although this is an existence proof (that is, under the appropriate conditions zero-crossings specify an image completely; see Appendix 2), there are probably more powerful ways of extending Logan's result to two dimensions. Their existence remains, however, an open question.

In its extreme form, our thesis may be summarized as follows. In order to construct a faithful representation of the image using only zero-crossings, it is necessary to filter it through a set of independent bandpass channels with about one octave bandwidth. Hence the masks (or receptive fields) that approximate the second directional derivative operator should, as closely as possible, be bandpass with about one octave bandwidth. (The exact maximum bandwidth depends on the input set and on eventual extra information: see later.) Such a system would allow the recovery of sharp intensity changes directly from the mask outputs, while providing the necessary basis for the recovery of the information contained in arbitrary intensity profiles. Logan's theorem does not provide any "reconstruction" method, to obtain the original signal from the zero crossings of its filtered versions. From the point of view of visual information processing, this is hardly a problem: There is clearly no need to reconstruct the original intensity values. The important point is that "discrete" symbols, like zero-crossing positions (and their sign), can represent with full information the original image and can be used for subsequent processing.

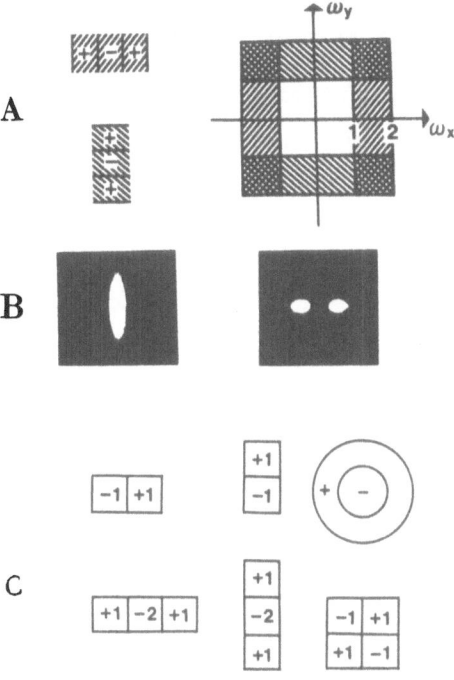

Figure 2:

(a) On the left are shown short bar-shaped masks at the vertical and horizontal orientations, and on the right, the amplitude of their (idealized) transfer functions. The bandwidth shown here is one octave, the maximum value for which Logan's theorem applies. (In practise, an ideal one-octave bandwidth requires side-lobes in the "receptive field"). If for each mask, zero-crossings are found along scan-lines lying perpendicular to the mask's orientation, these zero-crossings contain full information about that part of the image whose spectrum falls within the shaded region (on the right) of the Fourier plane. The remaining regions of the Fourier plane can be covered by similar masks of different sizes. Provided that there is sufficient overlap in the Fourier domain, information from different masks can in principle be combined to give the original image up to a single scaling factor.

(b) Shows an elongated mask, whose crossection is the difference of two Gaussians, together with its Fourier transform. Interestingly, if one uses masks constructed from the difference of two Gaussian curves (Wilson and Gieze, 1977) their Fourier transforms behave like ω^2 (Campbell and Robson, 1968) for values of ω that are small compared to σ. In other words, they approximate a second derivative operator.

(c) Such masks can also be regarded as approximating differential operators at a particular scale. The first-order operators $\frac{d}{dx}$ and $\frac{d}{dy}$ are straightforward, and the Laplacian center-surround operator is familiar from retinal ganglion cell studies. The second derivative $\frac{d^2}{dx^2}$ can be thought of as being made from the differences of two adjacent measurements of $\frac{d}{dx}$ and similarly for $\frac{d^2}{dy^2}$. The final operator $\frac{d^2}{dxdy}$ can be thought of as measuring the change in $\frac{d}{dy}$ over a small distance in the x-direction, or viceversa.

According to the point of view we have outlined above, the channels are not performing Fourier analysis, and the "bar-detectors" are not detecting bars. Their function is to measure the second directional derivative of intensity, and their receptive field geometry is chosen to ensure that their zero-crossings alone are very rich in information.

We should mention here that this rather direct way of locating zero crossings is not the only nor necessarily the best method from the point of view of computational efficiency. It can be shown that under certain conditions, the zeros in an image filtered through a concentric-type field provide an equivalent way of locating edges, whose orientation must then be represented. This suggests that zero-crossings in an image filtered through concentric receptive fields may also contain the whole information of the image. As we mentioned, the extension of Logan's theorem to this case in not yet available.

What experimental evidence is there that our thesis is relevant to biological visual systems? As we mentioned earlier, Logan's free and simple zero condition will almost always be satisfied in practice. The critical condition concerns the bandwidth. There is ample evidence for the existence in the human visual system of independent, spatial-frequency-tuned bandpass channels, of about one octave bandwidth. Precise estimates of the bandwidth vary considerably, however, ranging from very narrow (0.5 octaves) (Sachs and Robson, 1971) to very large values (Kulikowski and King-Smith, 1973, and Shapley and Tolhurst, 1973). More recent approaches based on spatial probability summation allow most of the existing psychophysical data to be fitted using medium bandwidth channels. The especially convincing estimates of Wilson and Gieze (1977) hover around an octave and a half (Legge, 1978). There is unfortunately little available information about channel characteristics in their normal (suprathreshold) conditions, although there are hints that their bandwidth may be somewhat narrower (Cowan, 1977). In any case, the channels are not the ideal one-octave bandpass filters that Logan's theorem, in its strict form, requires.

The important point is that Logan's theorem shows that zero-crossings of a 1 octave bandpass signal contains complete information, and the evidence is that they remain rich in information even when the 1-octave condition is relaxed, although it should be stressed that Logan's result cannot be extrapolated with abandon to "almost bandpass" functions. Experiments with 1.5 octave ideal bandpass Fourier polynomials with coefficients randomly chosen from a uniform distribution (on [-1, 1], indicate that only in 8% of the cases there are there insufficient zero-crossings to determine the signal (see Appendix 1). For ergodic bandpass Gaussian processes, Rice's formula shows that Logan's conclusion holds up to 1.67 octave (H.K. Nishihara, personal communication). Alternatively, we conjecture that the allowed bandwidth may be increased by adding extra information, for example the derivative of the signal at the zero-crossings. Clearly extra information of low-pass type must also be preserved by early visual information processing (for instance, rough measurements of the absolute intensity). In any case it becomes of considerable interest to determine the channel bandwidths under suprathreshold conditions.

Finally, observe that the physiological detection of zero-crossings need not depend on the detection of cells with zero response. For instance, near an intensity edge the zero-crossing in the bandpass signal is flanked by two peaks of opposite sign. Detection of zero-crossings can thus be performed on the basis of peaks, rather than zero responses.

Appendix 1

Logan's results apply to $B_\infty(\lambda)$ functions, i.e., the restrictions to the real line of entire functions of exponential type λ whose growth (on (R)) is less than exponential. In particular, they apply to periodic functions with the exception of theorem 4 (Logan, 1977), which can be specialized to periodic functions (Logan, personal communication). If we restrict ourselves to trigonometric polynomials, it is possible to illustrate Logan's results in a simple way. It should be stressed, however, that trigonometric polynomials are a very special case and in general erroneous inferences can be made from their special properties. With this "caveat" in mind, let us consider the real band limited function

$$h(t) = \sum_{-N}^{N} C_n e^{int} \qquad C_n = \overline{C}_{-n} \tag{1}$$

which can be extended to the complex plane as

$$h(z) = \sum_{-N}^{N} C_n e^{inz}$$

$h(z)$ is for instance bandpass with one octave bandwidth if

$$C_n \simeq 0 \qquad |n| \leq \frac{N}{2}$$

The complex free zeros of $h(z)$ are the complex zeros of $h(z)$ in common with its Hilbert transform $\hat{h}(z)$ where

$$\hat{h}(z) = \sum_{-N}^{N} \hat{C}_n e^{inz} \qquad \hat{C}_n = -i \ sign(n)C_n \tag{2}$$

Let us define, given $h(z)$

$$P(z) = \sum_{A+1}^{N} C_n e^{inz}$$

$$N(z) = \sum_{-N}^{-(A+1)} C_n e^{inz} \tag{3}$$

where A is the low-frequency boundary of the spectrum of $h(z)$ (assumed in the following bandpass).

Then the free zeros of $h(z)$ are completely characterized by the following three equivalent formulations:

The free zeros of $h(z)$ are such z^*:

$$P(z^*) = 0 \qquad N(z^*) = 0 \tag{a}$$

$$h(z^*) = 0 \qquad P(z^*) = 0 \tag{b}$$

$$P(z^*) = 0 \qquad P(\overline{z^*}) = 0 \tag{c}$$

Observe that if z is a zero, $\overline{z}$ is also a zero of $h(z)$; and if z is a zero, $z + 2k\pi k$ an integer, is also a zero.

The coefficients C_n of $h(z)$ may be determined by the $2N$ roots of $h(z)$ as the solutions of the system of $2N$ equations

$$\sum_{-N}^{N} C_n e^{inz_1} = 0$$

$$\ldots$$

$$\sum_{-N}^{N} C_n e^{inz_{2N}} = 0 \tag{4}$$

Let us now rewrite

$$h(z) = \sum_{-N}^{N} C_N e^{inz}$$

as

$$h(\varsigma) = \left(\sum_{0}^{2N} g_n \varsigma^n \varsigma^N \right) \tag{5}$$

with

$$\varsigma = e^{iz}, \vartheta_n = C_{n-N}, \mathrm{R}[z] = [0, \pi], N =: 2M$$

Thus the nontrivial zeros of $h(z)$ coincide with the zeros of $\sum_0^{2N} \vartheta_n \varsigma^n$, that is, a polynomial of order $2N$. If the $2N$ roots ς would be known, it would be possible to write $2N$ equations in the $2N + 1$ real unknowns (C_n):

$$\sum_0^{2N} \vartheta_n \varsigma_1^n = 0$$

$$\sum_0^{2N} \vartheta_n \varsigma_{2N}^n = 0 \qquad (6)$$

with

$$\varsigma = e^{iz}$$

Since the determinant of the roots is a Vandermonde determinant, it has always maximum rank if the roots are distinct. The question is under which conditions the real roots alone determine, apart a multiplicative constant, the set of C_n, i.e. $h(z)$. Clearly, multiple zeros, in particular multiple real zeros, cannot be allowed. Observe that if more than $2N$ real zero-crossings would be available (in a basic period) then $h = 0$.

Under the bandpass condition ($C_n = 0$ for $n \leq A$) there are at least $2A$ real zero-crossings per period. The real unknowns are $2b$, $b = N - A$, that is the number of non-zero C_n between N and A, counted twice because they are complex numbers. A sufficient condition to ensure that there are enough zero-crossings, and thus equations, is $A = M = \frac{N}{2}$, i.e., non-zero C_n (for $n > 0$) in $[M, 2M]$. Notice that $[M, 2M]$ i.e., one octave bandwidth would not be sufficient: in this case there would be at least $2M$ real roots but $2(M + 1)$ unknowns C_n. The matrix associated to the homogeneous equation in the "roots"

$$\begin{pmatrix} e^{-i2Mt_1} & e^{-i(A+1)t_1} & e^{i(A+1)t_1} & e^{i2Mt_1} \\ \cdots & \cdots & \cdots & \cdots \\ e^{-i2Mt_{2M}} & \cdots & \cdots & \cdots \end{pmatrix}$$

has rank at most $2M - 1$ (since there exists C_n such that $\sum C_n e^{inx}$ vanishes identically for $x = t_1 \ldots t_{2M}$) and this would just not suffice to specify the C_n modulus a multiplicative constant.

Although the less-than-1 octave condition is sufficient to ensure enough zero crossings, it is by no means necessary. In fact, there are classes of bandpass signals with a larger bandwidth and still enough zero-crossings.

In any case, even when there is a sufficient number of zerocrossings, the question still remains of whether the determinant of the matrix of the "roots" $|e^{intz}|$ has maximum rank $(2M - 1)$ and therefore the C_n can be determined (modulus a multiplicative constant). If the rank is less than $2M - 1$ then the C_n are not uniquely determined and as a consequence $h(z)$ is not determined by its real roots. Logan (1977 and personal communication) has proved that

a) if a free zero exist then $h(z)$ is not uniquely determined by its real roots and

b) if there are no free zeros, $h(z)$, provided its bandwidth is appropriate, is determined, modulus a multiplicative constant, by its real zerocrossings.

In the following, I will outline Logan's main theorems for the case of trigonometric polynomials.

Theorem 1

If $h(z)$ has 1 or more free zeros, the rank r of the determinant of the roots is $r < 2M - 1$.

Proof

$h(t)$ can be written as

$$h(t) = P(t) + N(t)$$

$$= e^{-i2Mt}\{\sum_0^{M-} g_n e^{int}\} + e^{i(M+1)t}\{\sum_0^{M-1} P_n e^{int}\} \tag{8}$$

$$= e^{-i2Mt} \prod^{M-1} (e^{it} - e^{i\delta_J}) + e^{i(M+1)} + \prod^{M-1} (e^{it} - e^{i\delta_J})$$

If ϵ is a free zero of $h(t)$ then I can divide $h(t)$ by the real function

$$f(t) = (e^{it} - e^{i\epsilon})(e^{it} - e^{i\bar{\epsilon}}) = (2ie^{\frac{it+\epsilon}{2}}\sin\frac{t-\epsilon}{2})(2ie^{\frac{it+\bar{\epsilon}}{2}}\sin\frac{t-\bar{\epsilon}}{2}) = A\sin\frac{t-\epsilon}{2}\sin\frac{t-\bar{\epsilon}}{2} \tag{9}$$

with A real.

The resulting $\frac{h(t)}{f(t)}$ is still a periodic bandpass function of the form

$$\frac{h(t)}{f(t)} = \sum_{-2M}^{-M} S_n e^{int} + \sum_{M}^{2M} S_n e^{int} \tag{10}$$

and actually of reduced bandwidth. Multiplication of $\frac{h(t)}{f(t)}$ by any arbitrary $[a - \cos(t - \sigma)], a > 1$ which can be always written as $C\sin\frac{t-\tau}{2}\sin\frac{t-\bar{\tau}}{2}$, provides a periodic bandpass function with the same bandwidth as the original $h(t)$ but different from it despite the same real zeros. Notice that if ϵ is not a free zero, $\frac{h(t)}{f(t)}$ will not be any more a periodic bandpass function. This means that the determinant associated with the homogeneous equation 7 has at most rank $r = 2M - 2$.

Theorem 2

If $h(t)$ has no multiple and no free zeros the rank of the determinant of the real "roots" is $r = 2M - 1$.

Proof

Clearly r cannot be $r > 2M - 1$. If h_1 and h_2 have the same bandwidth and the same real zeros, then

$$h_1 h_2 + \hat{h}_1 \hat{h}_2 = \sum_0^{2M-1} g_n e^{int} \tag{11}$$

$$h_1 h_2 - \hat{h}_1 \hat{h}_2 = \sum_0^{2M-1} P_n e^{int} \tag{12}$$

as it is easy to check by substitution of equation (2). If the real zeros are $2M$ and distinct, the Vandermonde determinant associated to the real roots of equation 12 is different from zero; thus, the unknowns g_n are identically zero. The same argument implies that all P_n are also identically zero. Thus, $\frac{h_1}{h_1} - \frac{h_2}{h_2} = M(t)$.

Now $M(t)$ is any function with the same zeros (real and complex) of h_1. But h_1 is a bandlimited function $h_1(t) = \sum_{-2M}^{2M} C_n e^{int}$ which is uniquely determined (apart from a multiplicative constant) by its 4M real and complex zeros. (See argument at pg. 11-12.) Thus h_1 and h_2 must coincide identically and the theorem follows. The theorem can be generalized allowing for real zeros.

Finally, a short remark about the multiple and free zero condition. It is rather intuitive that multiple and free zeros are not generic; assume, for instance, that the polynomial $\sum_{-N}^{N} C_n e^{int}$ has a free zero. It's enough to perturb one of the coefficients C_n to annihilate the free zero. Similarly, if the trigonometric polynomial is a sample function of a random process, the coefficients C_n would be random numbers, as well as the zeros of the associated polynomial $\prod^{2N}(\varsigma - \varsigma_i)$. The probability that a zero is free (i.e. with $\varsigma_i = \rho \; e^{i\theta}$, ς_i is free iff $\frac{1}{\rho} e^{i\theta}$ is also a zero) is usually very low.

Appendix II. (by H.K.Nishihara)

Logan's result can be extended to the case of a two-dimensional entire function $f(x, y)$ if it is bandpass in x with a band-width strictly less than an octave and band-limited in y (see figure 2a). In this case, the restriction of f to a one-dimensional line l_x in the x, y plane parallel to the x axis will be bandpass with less than an octave band-width. Provided the free-zero condition is met, Logan's theorem tells us that the zeros of f along l_x determine f there up to a multiplicative constant. To determine f everywhere up to a multiplicative constant, these parallel slices must be tied together.

The following lemma shows that Logan's theorem can be invoked for f restricted to a line l_θ which is not parallel to the X axis. l_θ will intersect all slices l_x parallel to the x axis, so determining f up to a multiplicative constant on l_θ determines f up to the same constant along each of the slices l_x.

Lemma

If $f(x, y)$ is ideally bandpass with band-width strictly less than an octave in x and band-limited in y then there is an $\epsilon > 0$ such that f along all slices, l_θ which make an angle $\theta < \epsilon$ with the X axis, will be bandpass with band-width less than an octave.

Proof

The support of the Fourier transform of f is confined in ω_x to the intervals $I_1 = (-2a + \delta, -a - \delta)$ and $I_2 = (a + \delta, 2a - \delta)$ and in ω_y to the interval $J = (-b, b)$ for some positive δ, a, and b. Observe that the support of the Fourier transform of a slice l through f is confined to the projection of the support of the Fourier transform of f onto the ω_l axis. The rectangles $I_1 \times J$ and $I_2 \times J$ will project into the intervals $(-2a, a)$ and $(a, 2a)$ on l_ω provided that l makes a sufficiently small angle with the x axis.

References

Campbell, F.W. and Robson, J.G., "Application of Fourier analysis to the visibility of gratings.," *J. Physiol. (Lond)* 197 (1968), 551-556.

Cowan, J.D., "Some remarks on channel bandwidths for visual contrast detection.," *Neurosciences Res. Prog. Bull.* 15 (1977), 551-556, fig. A12.

Hubel, D.H. and Wiesel, T.N., "Receptive fields, binocular interaction and functional architecture in the cat's visual cortex.," *J. Physiol. (Lond)* 160 (1962), 106-154.

Kulikowski, J.J. and King-Smith, P.E., "Spatial arrangement of line, edge and grating detectors revealed by subthreshold summation.," *Vision Res.* 13 (1973), 1455-1478.

Legge, G.E., "Sustained and transient mechanisms in human vision: temporal and spatial properties.," *Vision Res.* 19 (1978), 69-81.

Logan, B.F., Jr., "Information in the zero-crossings of bandpass signals.," *Bell System Tech. J.* 56 (1977), 487-510.

Marr, D., "Early processing of visual information.," *Phil. Trans. R. Soc. B. (Lond)* 275 (1976), 483-524.

Marr, D. and Poggio, T., "Theory of human stereopsis," *J. Opt. Soc. Amer.* 67 (1977), 1400.

Marr, D. and Poggio, T., "A computational theory of human stereo vision.," *Proc. Roy. Soc. B.(Lond.)* 204 (1979), 301-328.

Sachs, M.B., Nachmias, J. and Robson, J.G., "Spatial-frequency channels in human vision.," *J. Opt. Soc. Amer.* 61 (1971), 1176-1186.

Shapley, R.M. and Tolhurst, D.J., "Edge detectors in human vision.," *J. Physiol. (Lond)* 229 (1973), 165-183.

Wilson, H.R. and Gieze, S.C., "Threshold visibility of frequency gradient patterns.," *Vision Res.* 17 (1977), 1177-1190.

FEATURE DETECTORS, VISUMOTOR COORDINATION AND EFFERENT CONTROL [1]

Michael A. Arbib

Department of Computer and Information Science; and
Center for Systems Neuroscience
University of Massachusetts

I. A Brain Theory Perspective

The perspective of this paper on feature detectors is that of a neuroscientist who is not an experimentalist but, rather, a brain theorist. As was brought out in David Marr's talk, there are two complementary approaches to brain theory. The "bottom-up" approach starts from detailed information about individual neurons and tries to analyze their behavior when connected together in simple or highly regular neural nets. The "top-down" approach draws many ideas from AI (Artificial Intelligence) without in any sense being driven by it, in trying to analyze what pattern of interactions between what sort of sub-systems might be required to achieve some overall process. This talk will exhibit the interaction between the attempt to think about necessary processes in visually guided behavior and the data from neurophysiology concerning nerve circuits that subserve these tasks. While my own work is in theory, experiments will be continually coloring the story as we go along.

Within this framework, our interest in features can be defined in two complementary but related ways:

1) What are good descriptors for neurons in sensory systems? Without arguing whether a neuron is a feature detector or a feature receptor, we ask whether there is some useful way in which we can describe the behavior of neurons in at least the peripheral visual system.

2) The complementary "top-bottom" question is: what are good data structures in which to represent the results of processing as we go further and further from raw visual input to a representation of a scene useful for guiding the action of the organism.

The concern with this sort of description of cells in the visual system starts with Barlow's (1953) observation of "bug detectors" in frogs, but really enters the mainstream of neuroscience with such papers as Lettvin et al. (1959) and Hubel and Wiesel (1960). The transition to visually-guided behavior comes when we go beyond asking "What does the frog's eye tell the frog's brain?" or "What are the properties of cells in area 17?" to asking, for example, "What does the frog's eye tell the frog?" It's one thing for an intelligent human being to stick a microelectrode into a cell and then describe correlations between the antics in front of the animal and the activity of the cell (the notion of a "candidate code", Perkel and Bullock, 1968); it's another thing to say that these correlations are used by the brain that lies beyond that cell in actually guiding the behavior of the animal. My suggestion will be that, in looking at the growing data on visuo-motor coordination in frog and toad, we have some chance of understanding neural circuits that use various kinds of visual features and integrate them to guide the behavior of the organism.

Before turning to this discussion of visually-guided behavior, I want to discuss a problem that seems ignored in most discussions of feature detectors -- namely how large their receptive fields can be (McIlwain, 1967). In superior colliculus we see cells with receptive fields even larger than 50 degrees of visual angle. In his contribution to this volume, Horace Barlow has noted that our vernier acuity is much better than one would expect from the receptive field size of even foveal cells. Much of the machine vision literature assumes that visual information is kept punctate enough to provide precisely localized information about visual features. The large-window/small-window account of stereopsis given by Marr and Poggio in their contributions gives one account of how the brain's receptive fields might interact in a useful fashion. Another useful account comes from the work of Erickson (1974) on population coding, which shows that we could resolve some of these apparent problems with huge receptive fields if we stress not what a single neuron is doing but what a population of neurons is doing, so that peaks in intensity in a population, rather than a single neuron's activity, are seen as most important.

This returns us to the whole question of how the neural code is to be read out. If we imagine that each percept or plan of action requires a single cell somewhere that "puts it all together," it is hard to see what role populations could play. But if we think of actions mediated by pools of motor neurons played upon by other pools of neurons which in turn interact with a pool of visual system neurons, we come to stress the interaction of peaks of activation or ensemble averages in pools of neurons and there is no need to convert from a population code into a localized signal and then back again.

In addition to discussing the brains of amphibia, I will address a topic that is currently more AI rather than brain theory. This is the computation of optic flow--the vector field of movement of texture on the retina which occurs when an organism moves through the environment and which can convey information valuable to the organism as it moves. Our discussion of optic flow will not be at the level of individual neurons, but it will be in terms of layered computation. With this mention of layered computation, let me comment on how one makes the brain "manageable." The brain is a big place, and so we refer to anatomical regions, but these are so big in neural terms that it is not always a fruitful form of analysis to jump right down to the individual neuron immediately from the brain region. One useful approach represents brain regions as layered structures and then tries to understand how patterns of activity played out over different layers can interact in determining overall behavior. Another way of decomposing a region is into modules, where the stress is on units of vertical organization whose distributed interactions determine the overall behavior of the region (Szentagothai and Arbib, 1975; Mountcastle, 1978; Szentagothai, 1978).

2. The Amphibian Visual System.

Before discussing the amphibian visual system, I want to mention the work of Vernon Mountcastle (1976) on neural command functions for selective attention in the monkey. He and his colleagues record from cells in the parietal lobe of the monkey and can describe the responses of many of these cells in terms of the animal's intentions, but not in terms of the exact position of a stimulus or the exact characteristics of a motion. Such experiments suggest the way in which the type of methodology that I am talking about at the level of the amphibian will yield more and more information about the brains of mammals and even primates.

It is well known that if we monitor ganglion cells in the frog retina (Lettvin et al., 1959), we find them tuned to certain "things in the world" such as "bugs" and "enemies." However, if we look more subtly at the way in which the animal responds and try to describe in some detail what stimuli are most effective, it soon becomes clear that just looking at the group II ganglion cells of the frog retina, the so-called "bug detectors," is not enough. The activity in this population of cells alone does not allow prediction of the maximal behavioral response of the frog. Thus, even though it is attractive to refer to the group II cells as bug detectors, they are not bug detectors in the strictest sense. I now want to discuss competition between different stimuli and between properties of a stimulus in the amphibian visual system. Ingle (1968) observed that if a frog were confronted with two fly-like stimuli, it would often "choose" and snap at one of them. But in some circumstances it would choose neither, and in a few circumstances it would choose the "average fly."

Let us distinguish between brain theory and AI here. If you ask a computer scientist to come up with a program which, given a list of three locations, say, and the level of "bugness" at each of those, has to pick the maximum, he would find the task trivial. The real interest for the brain theorist is how to play out the prey-selection over a plausible neural circuit where no neuron is a centralized controller. The scheme we came up with (Didday, 1970, 1976) uses one layer of cells signalling some weighted "foodness" evaluation, and another layer which samples for each locale how much activity is going on elsewhere. To the extent that there is much activity going on elsewhere, they turn down the local "foodness" estimates. On this basis we came up with circuitry (plausible in terms of what we knew of the data in 1970) which could make a choice in a purely distributed way and in general would converge to direct the animal to snap at the most powerful stimulus, but would sometimes deadlock where two stimuli had turned each other down, and so neither could escape the competition with the other to take command of the system.

We have learned much since then that makes the problem more interesting. We now have much better quantitive data on amphibian behavior, so a challenge for modeling is to look at some of these data and try to take them into account. Ewert (1976), for example, has moved rectangles across the visual field of a toad and then used the frequency with which the toad responds as a measure of potency. If the rectangle is lengthened in the direction of movement -- "a worm" -- it becomes a more potent stimulus, but if it is elongated normal to the direction of stimulus--the so-called "anti-worm", the animal will be inhibited from responding. The toad without PT (pretectum and thalamus), does not make this discrimination -- the bigger the stimulus, no matter in which way, the more the animal will snap. This suggests the presence of a "worm map" and an "anti-worm map," with the anti-worm map located in PT and inhibiting the the worm map. The PT-leisoned frog (or toad) will snap at anything that moves.

In the amphibian brain, then, we have the competition between two worms or two flies for the attention of the animal, but we also have discrimination of whether to snap at or jump away from a stimulus. In current work (Arbib, Cromarty and Lara, to appear), we model these tectal-pretectal interactions to match quantified behavior data using circuits constrained by detailed neuroanatomy (Szekely and Lazar, 1976). This new model also addresses the facilitation phenomena observed by Ingle (1968, 1976).

We see, then, that the visuomotor system of the amphibian is a preparation in which we can begin to look at the problem of feature detectors not in a purely sensory mode, but with a concern for how these features are combined in determining action. And even in the frog visual system--that we thought was simple until we had 20 years of detailed study of it--we see how subtle is the integration of visual features in determining action. Many different regions are involved, and current stimuli must be integrated with short-term memory effects, such as habituation and facilitation, in determining the overall course of action of the animal.

In another experiment, Ingle (1976) studied the behavior of frogs jumping away from a large moving object. If he placed a barrier in front of the frog's preferred direction of jumping, he found that the frog now jumped in a different direction so as to avoid the barrier as well as to escape the "predator." In the prey-catching and the pure predator-avoiding situations, we can pretend that the animal is driven purely by a local stimulus, after "choosing" to which stimulus to respond. But this description is no longer tenable when we interpose barriers, for we see that there are many static features of the environment that have to be integrated in determining the response. Even in what we once thought was a relatively simple creature, we have to start analyzing its spatial map(s) of the world. The animal's action is not determined by single features, but in terms of integrated patterns of those features.

3. Representing the Visual Input in Higher Organisms.

In describing the representation of visual input in the brain of humans or in general-purpose machine vision systems (see Hanson and Riseman, 1978, for an excellent review of this literature), we start with very local features and then invoke processes to determine the presence of edges, consistent patches of texture, etc., as part of the process of segmentation. The resultant segments are candidates for meaningful surfaces in the environment. However, experiments in machine vision show that segmentation is far from straightforward -- a surface may be segmented into many regions or two surfaces may "bleed" into a single segment. In any case, these low-level non-semantic processes represent the data in a point-by-point feature description region-by-region form suitable for one further inference. (David Marr has offered the 2 1/2-D sketch as one idea of what this intermediate representation might look like. The papers by Marr and Poggio in this volume offer exciting approaches to the segmentation problem, but it must be stressed that there are many competing theories in the brain theory and machine vision literatures.)

To summarize: an important intermediate level of representation of a natural scene seems to involve segments which can, with further processing, be broken down and aggregated into surfaces which can then be recognized as cohering into known objects. As a first cut, we can talk about low-level vision, which is that which yields reasonably satisfactory segmentation; and high-level vision which uses "semantics" (such as "if it's blue and at the top of the picture maybe it's the sky") to provide an interpretation of the scene. However, we shall shortly see that the one-way flow of information suggested by the low-level "to" high-level decomposition is misleading.

Given my bias towards sensorimotor concerns, I find it useful to analyze the scene in terms of information to guide action, rather than simply aiming for a labelled segmented image. Optic flow is the pattern of movement across the retina obtained as the organism moves relative to our environment.[2] J.J. Gibson (1955, 1958) observed that optic flow affords many inferences relevant to the organism's motion in the world that are "low-level" in that they do not rest on recognizing objects. The motion of texture elements across the retina supports the inference of the location of a collision, and of the time until contact (Lee, 1974). Feature processing for motion detection and computing patterns of low level motion of texture elements provides much information useful for navigation. While these "low-level" processes can help the organism avoid collisions, more subtle visuomotor tasks will, of course, require recognizing what objects there are, and calling upon knowledge of those objects. Much of this knowledge will be procedural, in terms of how to interact with an object.

These considerations remind us that there is no simple hierarchy proceeding from low-level local visual features to more and more abstract representations until finally all objects in the scene are elegantly represented in some standard way. Rather, the organism is interacting with the world and there will be different parts of the brain taking these mapped representations and processing them in different ways which must be coordinated in some fashion to yield to the animal's overall behavior. In general we have what we might want to call the action/perception (Neisser, 1976; Arbib, 1981). Examining a picture in just one fixation is an unusual form of visual perception. In our normal interaction with the world, we move around and often do not know what we encounter without some exploratory behavior, walking around an object or prodding it or talking to it quietly to elicit enough sensory information to integrate it into a dynamic, on-going high-level representation. I have suggested (Arbib, 1981) that we view this representation as an assemblage of schemas. Having grown up in a world of pictures and words, we often forget how much there is of this interactive process in preception. We should not be misled by our success, once we have become sophisticated perceivers, in recognizing objects from a single static presentation.

As another antidote to the notion of one-way flow of visual information, note that the projection back from visual cortex to lateral geniculate in mammals contains more fibers than project "up" from lateral geniculate. I find it attractive (and, at this time, very very speculative!) to suggest that as cortex tries to develop a "model of the world," it does not require the "lower" levels to send raw undigested information, but rather it shapes the computations in the lower levels to provide information of a crucial kind for currently developing hypotheses. (Didday and Arbib, 1974, make a related suggestion, about the role of cortical-collicular, rather than cortical-thalamic, interactions.) Szentagothai and Arbib (1975) comment that the thalamic relays in auditory, visual and somato-sensory pathways exhibit a remarkable degree of structural similarity. In each case we have two kinds of cells: cortical "relay" cells, which receive inputs from specific sensory afferents, which they transmit as output onto the appropriate sensory cortex: and Golgi Type II interneurons. I want to stress the synaptic interaction between these two types. The sensory afferents synapse with the "relay" cell as well as with the interneuron dendrites; the interneuron dendrites form dendro-dendritic synapses with the "relay" itself; and there are triadic synapses which are in turn modulated by fibers coming down from the cortex. The whole thalamic structure then seems far removed from a mere relay, being very intimately "designed" in terms of taking account of what's going on up in cortex. (For some theoretical speculations concerning the possible role of cortical modulation of thalamus, see Harth, 1976; and see Grossberg, 1978, for the related, though somewhat ill-defined, notion of adaptive resonances.)

4. Looking Ahead.

One of the most perceptive discussions of feature detectors (prior to this symposium!) was offered by Barlow in 1972. Let me close by recalling Barlow's five dogmas from that paper:

1) "To understand nervous function, one needs to look at interactions at a cellular level rather than at either a more macroscopic or microscopic level, because behavior depends upon the organized pattern of these intercellular interactions." My emphasis on layers and modules is a partial but not a very strong disagreement with this; but I would suggest that much useful information is encoded in pattern of activity in neural populations.

2) "The sensory system is organized to achieve as complete a representation of the sensory stimulus as possible with the minimum number of active neurons." Jack Pettigrew commented at the meeting that he might spend a day trying to excite a cell only to have somebody come in with an appropriately ribbed sweater and activate the cell! This certainly seems consistent with the suggestion that a relatively small number of active neurons codes a visual stimulus. As Barlow was careful to emphasize in his 1972 paper, the idea of coding with a minimum number of active neurons does not entail the number is one, i.e., that there is a "grandmother cell". Rather he suggests that, perhaps in a thousand or two thousand cells, we might have the equivalent of a two thousand word description which is rich enough to give the brain all it needs to know about a figure, but coded in a way which puts it there ready for the brain to use.

3) "Trigger features of sensory neurons are matched to redundant patterns of stimulation by experience as well as by developmental processes."

4) "Perception corresponds to the activity of a small selection from the very numerous high-level neurons, each of which corresponds to a pattern of external events of the order of complexity of the events symbolized by a word." My reservation here is that those of us who talk about schemas (Arbib, 1981) as the units of representation might want to suggest that these units are played out over neural structures rather than single neurons. I do not have the faintest idea what sort of neural structure represents our knowledge of an apple!

5) "High impulse frequency in such neurons corresponds to high certainty that trigger feature is present." What our discussion of "model-building" as well as our discussion of the "false-match" problem in stereopsis and optic flow suggests is that cellular activity rarely codes simple presence of a feature, but rather is part of a pattern highly dependent on context as represented in the brain.

Footnotes

[1]This paper is an edited transcript of a talk presented on March 23, 1979. Subsequent research at the University of Massachusetts of Amherst and elsewhere has further contributed to the approach presented here. Preparation of this paper was supported in part by the National Institutes of Health under grant NS14971-01 from NINCDS.

[2]Computing the optic flow raises the same "false match" problems as arise in computing a disparity array in stereopsis. Marr and Poggio discuss stereopsis algorithms in their chapters. For studies on stereopsis, optic flow and related problems in neural net performance from my own group, see Amari and Arbib (1977), Arbib, Lawton, and Prager (to appear), Burt (1976), Dev (1976).

References

Arbib, M.A. 1980. Perceptual Structures and Distributed Motor Control,
    in V.B. Brooks, (Ed.),Handbook of Physiology, Section I:
    Neurophysiology,Vol. III. Motor Control, American Physiological
    Society.

Arbib, M.A., Cromesky, A.S. and Lara, R. (to appear). Modelling
    Studies of Visuomotor Coordination in Frog and Toad.

Arbib, M.A., Lawton, D.T., and Prager, J.M. (to appear). Computing the
    Optic Flow.

Barlow, H.B. 1972. Single units and Sensation: A neuron doctrine for
    perceptual psychology? Perception, 1, 371-394.

Burt, P.J. 1976. Stimulus Organizing Processes in Visual Perception,
    Ph.D. Thesis, Amherst: University of Massachusetts.

Dev, P. 1975. Computer simulation of a dynamic visual perception
    model, Int. J. Man-Mach. Stud., 7, 511-528.

Didday, R.L. 1970. The Simulation and Modelling of Distributed
    Information Processing in the Frog Visual System, Ph.D. Thesis,
    Stanford University.

Didday, R.L. 1976. A model of visuomotor mechanisms in the frog optic
    tectum, Math. Biosci., 30, 169-180.

Didday, R.L. and Arbib, M.A. 1975. Eye-movements and visual
    perception: a "two-visual system" model, Int. J. Man-Mach.
    Stud., 7, 547-569.

Erickson, R.P. 1974. Parallel "Population" Neural Coding in Feature
    Extraction, in F.O. Schmitt and F.G. Worden (eds.) The
    Neurosciences, Third Study Program, Mass.: MIT Press, 155-169.

Ewert, J.P. 1976. The visual system of the toad: behavioral and
    physiological studies on a pattern recognition system. In: K.V.
    Fite (ed.) The Amphibian Visual System, A Multidisciplinary
    Approach, New York: Academic Press, 141-202.

Gibson, J.J. 1955. The optical expansion-pattern in aerial location,
    Amer. J. Psychol., 68, 480-484.

Gibson, J.J. 1958. Visually controlled locomotion and visual
    orientation in animals, Br. J. Psychol., 49, 182-194.

Grossberg, S. 1978. A theory of human memory: self-organization and
    performance of sensory-motor codes, maps and plans. In:
    R. Rosen and F. Snell (eds.) Progress in Theoretical Biology,
    New York: Academic Press, 233-374.

Hanson, A.R. and Riseman, E.M. 1978. Computer Vision Systems, New
    York: Academic Press.

Harth, E. 1976. Visual Perception: A Dynamic Theory, Bird Cybernetics, 22, 169-180.

Hubel, D.H. and Wiesel, T.N. 1959. Receptive fields of single neurons in the cat's striatic cortex, J. Physiol., 148, 574-531.

Ingle, D. 1968. Visual releasers of prey-catching behavior in frogs and toads, Brain Behav. Evol., 1, 500-518.

Ingle, D. 1976. Spatial vision in anurans. In: The Amphibian Visual System, edited by K.V. Fite, New: Academic Press, 119-141.

Lee, D.N. 1974. Visual information during locomotion. In: R.B. McLeod and H.L. Pick, Jr. (eds.) Perception: Essays in Honor of J.J. Gibson, Ithaca, New York: Cornell University Press, 250-267.

Lettvin, J.Y., Maturana, H., McCulloch, W.S. and Pitts, W.H. 1959. What the frog's eye tells the frog's brain. Proc. IRE, 47, 1940-1951.

McIlwain, J.T. 1976. Large Receptive Fields and Spatial Transformations in the Visual System, International Review of Physiology, Neurophysiology II, 40, 223-248.

Mountcastle, V.B. 1976. The world around us: Neural command functions for selective attention. The F.O. Schmitt Lecture for 1975. Neurosci. Res. Program Bull., 14, suppl.1.

Mountcastle, V.B. 1978. An organizing principle for cerebral function: the vast module and the distributed system. In: G.M. Edplman and V.B. Mountcastle (eds.) The Mindful Brain, The MIT Press, 7-50.

Neisser, U. 1976. Cognition and Reality: Principles and Implications of Cognitive Psychology. San Francisco: W.H. Freeman.

Perkel, D.H. and Bullock, T.H. 1968. Neural Coding, Neurosciences Research Program Bulletin, 6, 3.

Szekely G. and Lazar, G. Cellular and Synaptic Architecture of the Optic Teclum. In: R. Llinas and W. Precht (eds.) Frog Nesobiology, 407-434.

Szentagothia, J. 1978. The neuron network of the cerebral cortex: a functional interpretation, Pec Roy Soc Lork, B 201, 219-248.

Szentagothis, J. and Arbib, M.A. 1975. Conceptual Models of Neural Organization, Mass.:The MIT Press.

FEATURE DETECTORS AND SPEECH PERCEPTION: A CRITICAL EVALUATION[1]

Joanne L. Miller

Northeastern University

and

Peter D. Eimas

Brown University

I. Introduction

The nature of human language and the processes involved in its production and comprehension have long been a concern to a variety of disciplines, including linguistics, psychology, and speech science. One reason surely is the fascination with language itself, a remarkable accomplishment that is at the same time universally shared by all human cultures and restricted to the human species.[2] Another reason for this great concern is the widely held belief that understanding the way in which language is processed will yield insights into the basic characteristics of human perception, cognition, and complex sequences of action.

One aspect of language that has received intensive study is speech--the vehicle that carries the linguistic message. Speech is a complex, continuously varying acoustic signal that is generated, _via_ equally complex sets of rules, by the speaker in order to convey a particular linguistic message. When a listener attends to speech, he or she perceives (although not always at the level of conscious awareness) a sequence of distinct and ordered units, the phonetic segments of the language--roughly the consonants and vowels.[3] This sequence of perceived phonetic segments, in conjunction with knowledge of the phonological, syntactic, and semantic rules of the language, allows the listener to derive the intended message of the speaker. A critical component of language comprehension, then, is the initial analysis of the waveform which yields a sequence of ordered phonetic segments. Two primary issues regarding this stage of processing are (1) which characteristics or properties of speech signal carry information relevant to the identification of phonetic segments and (2) what is the nature of the perceptual mechanism that receives as its input the continuously varying acoustic signal and provides as its output the discrete phonetic segments.[4]

Many years of research on the perception and production of speech have yielded substantial information about the properties that distinguish phonetic segments. Although we are far from being able to write a complete description of the relation between acoustic signal and perceived phone, research has provided sufficient knowledge to permit computer-controlled synthesis of highly intelligible, if not wholly natural, speech (e.g., Mattingly, 1968). In addition, this research has shown that the relation between speech and perceived phone is quite complex (e.g., Liberman, Cooper, Shankweiler, and Studdert-Kennedy, 1967). First, it is not the case that stretches of the acoustic signal, acoustic segments, correspond in a one-to-one fashion with the perceived phonetic segments. Rather, there is both a one-to-many and a many-to-one mapping, in that one acoustic property may simultaneously provide information about a number of phonetic segments and, conversely, the information for one phonetic segment may be spread across a number of acoustic segments. Second, the acoustic properties that specify phonetic distinctions appear not to be invariant, but rather to vary systematically in their form as a function of such factors as speaker, phonetic context, and speech tempo. This apparent lack of invariance requires that a major goal of any theory of speech perception is to rationalize the perceptual constancy that does in fact exist; that is, it must explain how the listener perceives the same phonetic segment when it is specified by a variety of context-conditioned cues.

Attempts to solve the problem of perceptual constancy fall into two major classes. First, a number of investigators have sought to discover properties of the signal, usually higher-order relational characteristics, that can be considered invariant cues for a given phonetic distinction (e.g., Cole and Scott, 1974; Lisker and Abramson, 1964; Stevens, 1975; Stevens and Blumstein, 1981). Although these invariant cues have proved to be extremely elusive, there has been some progress (Blumstein and Stevens, 1979, 1980; Stevens and Blumstein, 1978, 1981). To the extent that this strategy ultimately proves successful, the primary requirement of the perceptual mechanism is that it is capable of processing these attributes, which are often relational in nature. The second approach to the problem of perceptual constancy has been based on the assumption that the perceptual mechanisms, themselves, perform the necessary operations to provide a constant percept. Such models include the motor-theoretic models of Liberman and his colleagues (Liberman, Cooper, Harris, MacNeilage, 1963; Liberman et al., 1967), the analysis-by-synthesis models of Stevens and his associates (Stevens, 1960; Stevens and House, 1972), and some, but by no means all, feature detector models (e.g., Eimas and Corbit, 1973), which are the focus of this discussion. According to these models, constancy is provided by the perceptual system of the listener--it is not inherent in the signal, at any level of description. It is important to emphasize that in neither of the two major approaches to the problem of invariance can the description of the acoustic properties and the perceptual mechanism be considered independently. The nature of the information assumed to be phonetically relevant constrains the class of possible perceptual mechanisms, and at the same time the nature of the postulated mechanism constrains the nature of the acoustic information that can be assumed to signal a linguistic distinction. Thus, as in communication systems of other species (e.g., Hoy, 1974; Hoy, Hahn, and Paul, 1977; Newman, 1977), there is an intimate tie between the communicative signal that this produced and the mechanism that has evolved for its perception.

The theoretical approach to speech perception based on feature detectors has received considerable attention over the past several years. One of the underlying assumptions of any feature detector model of perception is that perception involves at least two stages of processing. The first is an analytic stage, during which the signal is decomposed into a set of feature values or properties. This is presumably accomplished by a set of perceptual mechanisms-feature detectors--that in the ideal case respond invariantly to certain "trigger features" in the signal. The advantage of this system is that a large number of patterns can be described in terms of combinations of a small set of features. Consequently, it is only necessary to assume a small set of detectors, rather than a "template" for each perceptible pattern. During the second stage of processing, the features are read or recombined by some higher-level processor to form an integrated unit, which is then perceived (e.g., Blakemore, 1975; Neisser, 1967; Werner, 1974). Applied to speech perception, the argument is that the speech signal is analyzed initially in terms of features and that these features are subsequently unified to form the perceived phonetic segments.

The purpose of this paper is to evaluate the feasibility of feature detector models of speech perception. Our discussion is divided into four parts. In the first section, we examine some of the factors that supported the initial interest in developing and applying feature detector models to speech, focusing on those characteristics of speech that made the model seem particularly appropriate. In the next section, we turn to a discussion of the initial empirical support for the model, which was obtained with a modification of the selective adaptation procedure that set the stage for numerous studies aimed at testing and refining the model. In the third and fourth sections we consider some of the more serious shortcomings of the model that were revealed by subsequent research, especially with respect to the ability of the model to account for perceptual constancy across contextual variation. Finally, we conclude by reconsidering the possible role of feature detectors in the total process of speech perception. For general reviews of speech perception, which provide background material for the issues to be discussed in this paper, the reader is referred to Darwin (1976), Pisoni (1978), and Studdert-Kennedy (1974, 1976). Reviews of the selective adaptation literature are found in Ades (1976), Cooper (1975), and Eimas and Miller (1978).

II. Feature detector models and speech perception: Background

The application of a feature detector model to speech perception was first seriously entertained in the early 1970's (Abbs and Sussman, 1971; Eimas and Corbit, 1973; Lieberman, 1970; Stevens, 1975), and has since been one of the dominant models of speech processing, although not without serious detractors (e.g., Diehl, 1981; Remez, 1979). The initial attractiveness of this model vis-a-vis speech was undoubtedly due to the interaction of a number of factors.

One of the most important of these was the relative success of this type of model in explaining pattern perception in other modalities, most notably visual patterns. A number of investigators, employing diverse techniques, had accumulated strong inferential evidence for the existence of channels of analysis (feature detectors[5]), sensitive to specific visual properties, and progress was also made in defining the operating characteristics of these mechanisms (see Anstis, 1975, Stromeyer, 1978, and Eimas and Miller, 1978, for reviews of this literature). Moreover, neurophysiological research with nonhuman species, especially the cat and monkey, provided considerable physiological evidence for this type of model, in that direct evidence was obtained for the existence of neurons selectively tuned to aspects of a visual stimulus (e.g., Blakemore (1975) for a review, and Hubel and Weisel (1962, 1965, 1968) for a selection of some of the early pioneering studies). It seemed reasonable, in light of this research, to assume that selectively tuned neural structures subserved aspects of speech perception in the human as well.

A second major development, which influenced the interest in feature detector models, was the rapidly accumulating evidence that in many nonhuman species, including insects, amphibians, birds, and mammals, there exist specialized mechanisms, some of which can be considered "feature detectors," for the analysis of biologically relevant signals, including those used for communication (see Newman, 1977, for a review of these findings). At the same time, linguistics and psychologists were proferring arguments for the existence of species-specific mechanisms for the processing of language and speech (Chomsky, 1972, 1975; Liberman, 1970; Liberman, et al., 1967; Studdert-Kennedy and Shankweiler, 1970), although the precise nature of these mechanisms was often left unspecified. The assumption that these mechanisms might take the form of neural analyzers, tuned to those features of the acoustic signal that were linguistically relevant and analogous to mechanisms found in other species, was extremely attractive, as well as theoretically parsimonious.

A feature detector model for speech perception was also appealing in that it provided an alternative to existing models of speech perception, which had proved to be inadequate mainly for two reasons. First, these models, including the motor theory of Liberman (e.g., Liberman, et al., 1967), the analysis-by-synthesis models of Stevens (e.g., Stevens and House, 1972), and the proposals of Fant (1960), were not explicit in their formulations, and hence difficult, although perhaps not impossible, to test empirically. The feature detector model, at least the version offered by Eimas and Corbit (1973), was sufficiently specific to be tested experimentally, and indeed considerable effort was expended with this purpose in mind. Second, there was growing evidence that very young infants, with minimal linguistic experience, possess highly developed capacities to perceive small variations in speech sounds and, moreover, they were shown to do this in what would come to be a linguistically relevant manner (e.g., Eimas, Siqueland, Jusczyk, and Vigorito, 1971; Moffitt, 1971; Morse, 1972; Trehub and Rabinovitch, 1972). That is to say, it was shown that infants discriminated two sounds that differed along an acoustic dimension when they were phonetically distinct (for the adult), but not when the acoustic difference failed to signal a phonetic distinction, even though the absolute acoustic difference between the two sounds to be discriminated was the same in each case. Thus, infants, before they are able to produce precisely controlled speech sounds, are able to perceive speech in terms of perceptual categories that closely correspond to the phonetic categories of adults. Data of this nature implicate a perceptual mechanism that is presumably a part of the biological endowment of the infant and operative without the necessity of relying on mediation by the production system or knowledge of higher levels of language. These latter factors are necessary for the successful application of most existing models of speech perception. A feature detector model in which the assumption was made that the detectors are tuned to the critical acoustic information in the signal and automatically classify the signal into phonetic features, thereby allowing for "direct perception" (cf. Gibson, 1979) seemed, at least initially, to be a viable solution to the problem.

A final factor initially suggesting the feasibility of a feature detector approach to speech has to do with the issue of the units of analysis. As we stated above, according to a feature detector model, the signal is initially analyzed into component features and then these features are later recombined to form the final representation corresponding to the unitary percept. An advantage of this approach is that a small set of defining features, in various combinations, can in principle specify an entire set of patterns. One problem for a detector model is how to arrive at this set of features. This is not obvious in the case of visual perception and substantial work has been invested in attempting to specify the features that best describe a set of visual patterns (e.g., Blakemore and Campbell, 1969; Gibson, Osser, Schiff, and Smith, 1963; cf. Neisser, 1967). However, in speech this was not a problem; the features had already been delineated by decades of linguistic analysis, which defined each phonetic segment in terms of a small number of phonetic features that distinguish the phonetic segments (of a given language) from each user. Although the precise feature system that best describes the segments is still a matter of debate, most students of speech would agree that, at least for purposes of phonological analysis, a featural description is considerably more parsimonious than a system based on holistic phonetic segments (see, for example, Chomsky and Halle, 1968).

To illustrate the advantage of a feature-based system we will consider the feature of voicing, which is included in virtually all feature systems. Phonetic segments in English can take on one of two voicing values, voiced (e.g., {b, d, g, z, l, r, a, o, i}) or voiceless (e.g., {p, t, k, s, f, ʃ , tʃ }). For many phonological operations in the language, it is considerably simpler to write the rules in terms of the feature voicing than in terms of phonetic segments. A good example is the operation that changes a regular verb into its past tense form, as when "walk" becomes "walk" +{t} (written in English orthography as "walked"), or when "grab" becomes "grab" + {d} (written as "grabbed"). What happens is that either the suffix {t} or {d} is added to the present tense form of the verb; but which suffix is added is quite systematic. If the word ends in a voiceless segment, as in the case of "walk," then a {t}, which is voiceless, is added. If, however, the words ends in a voiced segment, as in the case of "grab," then a {d}, which is also voiced, is added. In phonological terms, the two adjacent segments within the syllable take on the same voicing value. Whereas this rule can be described in a simple and elegant manner at a featured level of description, its specification in terms of phonetic segments is much more cumbersome.

Given the advantage of a feature-based system for linguistic analysis, a longstanding issue in speech perception has been whether these features have any "psychological reality"; that is, whether they serve as units during the processing of speech. By the time of the early 1970's, there was a variety of data that could be interpreted as evidence for a feature-based perceptual process and the prevailing belief was that phonetic features were in fact extracted during perception (e.g., Miller and Nicely, 1955; Studdert-Kennedy and Shankweiler, 1970; but see also Studdert-Kennedy, 1976; and Parker, 1977). This evidence includes, among others, results from listening tests conducted in noise and with certain kinds of acoustic filtering (e.g., Miller and Nicely, 1955), from judgements of similarity across speech sounds (e.g., Greenberg and Jenkins, 1964), and from numerous tests of the identification and discrimination of sounds that varied along a single feature (see Liberman, et al., 1967). Given the general acceptance of feature extraction as part of perception, much of the research on this issue was subsequently aimed at assessing whether features were extracted independently of one another whether there were interactions during processing, either at the stage of feature extraction or feature recombination (e.g., Eimas, Tartter, Miller, and Keuthen, 1978; Haggard, 1970; Smith, 1973).

Inasmuch as it was widely accepted that features were extracted during perception, the critical issue became that of specifying the nature of the mechanism that performed this operation. And feature detectors, as noted above, seemed a likely candidate. It is particularly important to emphasize that, at least implicit in some of the original formulations of a feature detector model for speech (e.g., Eimas and Corbit, 1973), was the assumption that the detectors were sensitive to linguistic features, not acoustic features or properties, which was in effect an attempt to resolve the lack of invariance between acoustic signal and perceived phone by assumption. Translated into the terminology of features, the problem is that a given phonetic feature (such as voicing) can be cued by a variety of acoustic properties or features (e.g., voice-onset-time, duration of silence, duration of preceding vowel) and that these cues are themselves context-dependent (see, for example, Eimas and Miller, 1978; Pisoni, 1978). Thus, if the feature detector were truly a phonetic feature detector, it would be able to integrate the many context-conditioned cues to yield as its output a single feature value (e.g., voiced or voiceless). In this way, the invariance problem would be solved; the detector would provide the invariance and, in doing so, would account for perceptual constancy. (Of course a remaining problem would be to specify the actual operating characteristics of the mechanism--the detectors--that performed this integration.) Once the phonetic features had been extracted from the signal, they (that is, the feature values themselves) were then available for integration into phonetic segments by higher levels of processing. As we will see in following sections, a significant portion of the research on feature detectors has been aimed at assessing whether in fact these mechanisms operate in terms of phonetic features or whether they serve instead to analyze specific acoustic properties, which would leave the problem of perceptual constancy unsolved.

III. Initial evidence for feature detectors for speech

Although the idea of applying a feature detector model to speech perception had been considered by a number of investigators (e.g., Abbs and Sussman, 1971; Stevens, 1975), the first empirical test of it was conducted by Eimas and Corbit (1973). Since their initial study forms the basis of nearly all subsequent research on the topic, it is important to describe their model and experimental technique in some detail.

Eimas and Corbit focused on the phonetic feature of voicing and, in particular, on one acoustic cue for voicing, voice-onset-time (VOT) (Abramson and Lisker, 1970; Lisker and Abramson, 1964). In syllable-initial consonants ({b, d, g, p, t, k}) VOT is defined in articulatory terms as the time between the release of the consonant and the onset of vocal fold vibration. In acoustic terms it is the time between the initial release burst of the consonant and the onset of periodicity in the acoustic waveform. In English, consonants with VOT values greater than approximately +30 msec are perceived as voiceless ({p, t, k}) whereas those with VOT values less than about +30 msec are perceived as voiced ({b, d, g}).[6] Eimas and Corbit proposed that the phonetic feature values, voiced (V+) or voiceless (V-), were assigned by a set of feature detectors such as those shown schematically (by the solid lines) in Figure 1.

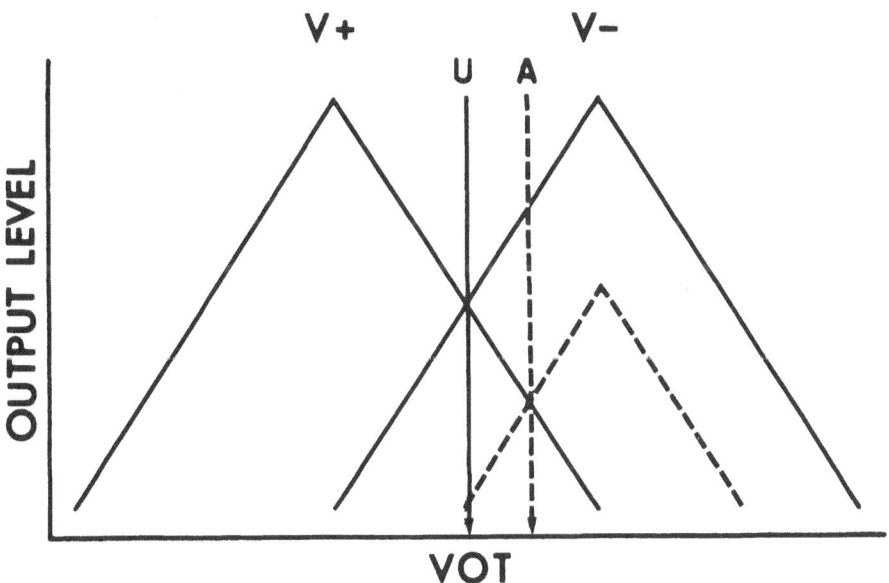

Figure 1: Model of a feature detector system. Output level is given as a function of input VOT value prior to adaptation (solid lines) and following adaptation (dashed lines) with a stimulus from the voiceless (V-) phonetic category.

Here we see two channels of analysis, one corresponding to the V+ value and the other corresponding to the V- value. The input to the

detector is assumed to be the auditory correlate of VOT and the output
is a binary decision, V+ or V-. Each channel is further assumed to be
sensitive to a range of VOT values, being maximally responsive at the
VOT value corresponding to the modal value in speech production. The
phonetic boundary between categories is located by assumption at that
VOT value to which the two channels are equally responsive. In
essence, the system functions by responding to a graded VOT signal and
yielding as its output a binary voicing decision, V+ or V-.

In order to test this model, Eimas and Corbit modified a procedure
of selective adaptation that had proved successful in examining the
channels of analysis for visual stimuli (e.g., Blakemore and Campbell,
1969; McCollough, 1965) as well as nonspeech auditory stimuli (e.g.,
Ward, 1973). The rationale for their procedure is as follows. If
listeners in an unadapted state identify a series of sounds that vary
in VOT from a value appropriate for a voiced consonant (e.g., {ba}) to
one appropriate for a voiceless consonant (e.g., {pa}), then estimates
of the category boundary between {ba} and {pa} as well as the point at
which the response curves from the two detectors presumably cross can
be obtained. If listeners are then exposed repeatedly to an exemplar
from one of the phonetic classes, say the voiceless class ({pa}), then
the voiceless detector should be fatigued and the output of that
channel should be markedly reduced, as shown by the dotted lines in
Figure 1. As a consequence of adaptation, the location of the
crossover point of the two response curves will be displaced toward
the value of the adapting stimulus and the identification function for
the same VOT series will have a boundary value likewise shifted toward
the adapted category. That is, some stimuli that were assigned the to-
be-adapted value before adaptation ({pa}) would be assigned the
unadapted value ({ba}) after adaptation. Such a shift in the
identification function would provide inferential evidence for the
existence of feature detector mechanisms of the sort proposed.

In their original experiment, Eimas and Corbit assessed the
perception, before and after adaptation, of two series of stimuli that
varied in VOT, one ranging from {ba} to {pa} and the other from {da}
to {ta}. These two series differed from each other on the feature of
place of articulation. This feature is defined in articulatory terms
by the locus of the major constriction in the vocal tract during the
production of the consonant. The consonants {b} and {p} are labial
consonants, produced with the major constriction at the lips, whereas
{d} and {t} are alveolar consonants, produced with the tongue at the

alveolar ridge behind the teeth. The distinction in place was cued acoustically by a change in the starting frequency, direction, and extent of the formant transitions, those regions of intense energy concentration corresponding to the resonances of the vocal tract that change in center frequency during the first 40–50 msec of the sound. Within each series, the stimuli varied only in VOT, cueing the phonetic distinction of voiced versus voiceless. Perception of both series was assessed before adaptation and after adaptation with the two voiced endpoint stimuli, {ba} and {da}, and the two voiceless endpoint stimuli, {pa} and {ta}. The results of the study are shown in Figure 2.

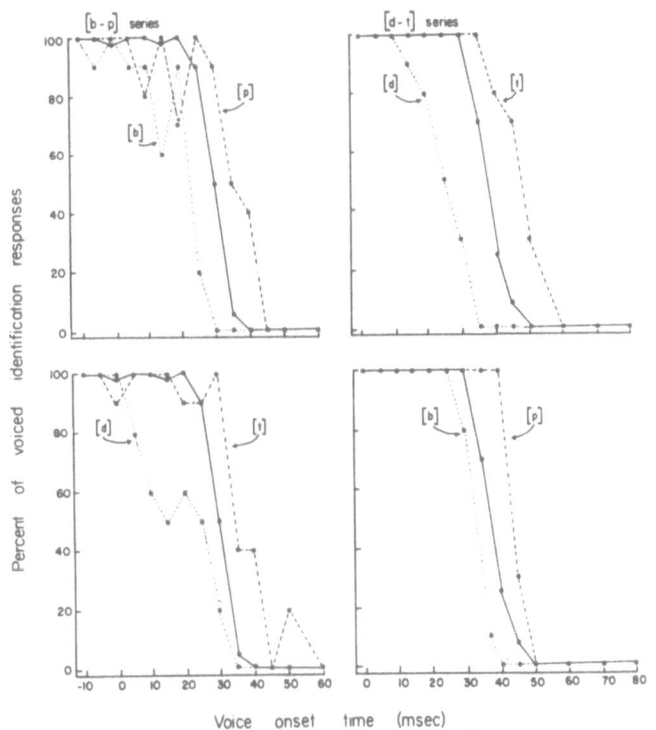

Figure 2: Percentages of voiced identification responses ({b} or {d}) obtained with and without adaptation for a single subject. The functions for the {b,p} series are on the left and those for the {d,t} series are on the right. The solid lines indicate the unadapted identification functions, and the dotted and dashed lines indicate the identification functions after adaptation. The phonetic symbols indicate the adapting stimulus. From Eimas and Corbit (1973).

First consider the effect of adapting with {ba} or {pa} on the {ba–pa} series and adapting with {da} or {ta} on the {da–ta} series. In all cases, adaptation shifted the identification function in the predicted direction, toward the voicing value of the adapting stimulus. Second, and of particular importance, was that analogous shifts were obtained when the adapting and test stimuli had different place of articulation values. Thus, adapting with {ba} and {pa} affected the perception of the {da-ta} series, and conversely, adapting with {da} and {ta} affected perception along the {ba-pa} series. These cross-series effects were taken as evidence that the adaptation effect was operating at the level of the phonetic feature, rather than at the level of the phonetic segment or syllable.

Following the lead of Eimas and Corbit, a number of investigators began testing various aspects of the model, typically using a selective adaptation technique very similar to that introduced in the original experiment. These investigations were not limited to the feature of voicing (as cued by VOT), but tested the generality of the model by applying it to other features, such as place of articulation (e.g., Cooper, 1974a; Cooper and Blumstein, 1974) and manner of articulation[7] (e.g., Diehl, 1976; Miller and Eimas, 1977). Among the many issues addressed by this research were the following : Is the site of adaptation (and hence the location of the detectors) at a peripheral or central level of processing (e.g., Ades, 1974a; Sawusch, 1977)? Is the response of the detector channel diminished by adaptation across its entire operating range or only near the phonetic boundary (e.g., Miller, 1975, 1977b)? Do adapting stimuli that themselves vary along the dimension under test produce differential magnitudes of adaptation, as implicitly assumed by the model (e.g., Anderson, 1975; Miller, 1977b)? What are the actual tuning curves of the detectors (e.g., Miller, 1977b)? Is the operation of the detectors language-specific (e.g., Foreit, 1977)?

There are two additional, closely related, issues that received substantial attention and that are probably the most critical in terms of the viability of the model. The first is whether the adaptation effect reflects an actual change in observer sensitivity resulting from the fatigue of a detector, as originally assumed, or whether it is the result of some form of contrast. The second is whether the detectors, as revealed by adaptation effect, are operating at an auditory or phonetic level of processing. These issues are considered in turn in the following two sections.

IV. Source of adaptation effect: Sensory fatigue versus contrast.

A primary assumption of the detector model originally proposed by Eimas and Corbit (1973) is that repeated exposure to a stimulus that activates a detector causes that detector to become less sensitive, that is, it is fatigued. The reduced output of the fatigued detector, in relation to the normal output of the unfatigued opposing detector, produces a shift in the crossover point of the response curves of the two detectors, resulting in the category boundary shifts typically measured in selective adaptation experiments (see Figure 1). According to this account, the source of the adaptation effect is sensory fatigue.

A number of early attempts were made to explore alternative explanations of the adaptation effect. For example, Ainsworth (1977) tested, and then rejected, an account based on the retuning of a detector's response characteristics. Similarly, Sawusch, Pisoni, and Cutting (1974) rejected an account based on alterations in response bias, and Cooper, Ebert, and Cole (1976), on the basis of a signal detection analysis, concluded that a major portion of the adaptation effect was indeed sensory in nature (but cf. Elman, 1979, for contradictory findings). On the basis of data such as these, as well as a number of additional arguments, Eimas and Miller (1978) concluded in a review of the issue that the available evidence supported an interpretation based on the fatigue of feature detectors.

More recently, the explanation of adaptation effects in terms of the fatigue of detectors has again been called into question (Diehl, 1981; Diehl, Elman, and McCusker, 1978; Diehl, Lang, and Parker, 1980; Elman, 1979; Simon and Studdert-Kennedy, 1978). The most extensive attempt to challenge the fatigue explanation has been that of Diehl and his colleagues. They propose that the boundary shifts seen in adaptation experiments are due to a contrast phenomenon, whereby the adapting stimulus, which is typically a good exemplar of one phonetic category, provides a reference against which the test stimuli are compared. Since the boundary stimuli are by definition poor exemplars of the category to which the adaptor belongs, they are assigned to the opposing category. This results in a shift in the category boundary toward the category of the adapting stimulus--exactly what is found in adaptation experiments. This account of the adaptation effect, which is derived from Helson's (1964, 1971) general theory of adaptation level (AL), has been termed the contrast hypothesis (Diehl et al.,

1980). It is important to point out, as has Diehl (1981), that the contrast effects predicted by AL theory could be operating at either a relatively early sensory/perceptual level or a later level of decision/response selection; there are debates within the AL literature itself as to the level of processing at which the contrast effect arises. Thus, the contrast hypothesis is not so much an attempt to explain adaptation effects in terms of decision, rather than sensory phenomena, but rather to account for these effects without postulating the fatiguing of detectors.

The strategy used by Diehl et al. (1978, 1980) to provide empirical support for the contrast hypothesis was to demonstrate that effects obtained using a selective adaptation paradigm can also be obtained with the simpler contrast paradigm. The essential difference between the two paradigms concerns the way in which the stimuli are presented. In a typical adaptation study, perception of test items is assessed before and after repeated exposure to an adapting stimulus. The direction and magnitude of the shift in category boundary as a function of such exposure, which reflects a change in the way in which boundary stimuli are identified, is taken as a measure of adaptation. In a contrast paradigm, a single context stimulus is presented before the test item, which is a stimulus from the boundary region. The change in identification of the test stimulus, as a function of the identity of the context stimulus, is taken as a measure of contrast. Since only a single presentation of the context stimulus precedes the test item, it is assumed that sensory adaptation--fatiguing of a detector--cannot be responsible for the effect. The argument is that to the extent to which both paradigms yield analogous results, the more reasonable and parsimonious it is to ascribe a single basis to both sets of effects, namely, the operation of contrast, as predicted by AL theory. In a series of studies, Diehl et al. (1978, 1980) demonstrated a number of parallels between the two paradigms; for example, effects analogous to cross-series adaptation effects originally found by Eimas and Corbit (1973), and discussed earlier in this paper, can be obtained with a contrast procedure. On the basis of the close correspondence between the results of adaptation and contrast studies, Diehl and his colleagues concluded that the effects obtained in selective adaptation experiments are a function of contrast.

The viability of the contrast hypothesis rests on the demonstration that the adaptation and contrast paradigms produce analogous results.

In a particularly clever experiment comparing the two procedures, Sawusch and Jusczyk (1981) showed that this is not always the case. The test stimuli for their adaptation study consisted of a {ba–pa} series whose members varied in VOT. The adaptors included a good {ba} from the series, a good {pa} from the series, and a {spa}. The {spa} was constructed, in accordance with its natural acoustic properties, by adding a segment of frication to the good {ba}. Thus, the {spa} had the same spectral configuration as the {ba} adaptor, but the same perceived phonetic quality as the {pa} adaptor. As expected, adaptation with {ba} shifted the category boundary toward the {ba} end of the series and {pa} had the opposite effect. Adapting with {spa} had the same effect as adapting with {ba}; that is to say, as an adaptor {spa} behaved in terms of its spectral properties, rather than its phonetic quality, a finding which agrees with the well-documented conclusion, discussed in the next section, that adaptation has an auditory, rather than phonetic, basis. In the companion contrast study Sawusch and Jusczyk (1981) assessed the identification of a boundary stimulus from the {ba–pa} series as a function of being paired with one of the three adaptors--{ba}, {pa}, and {spa}. As in the case of adaptation, when paired with {ba}, the ambiguous stimulus was heard more often as {pa}, and when paired with {pa}, it was heard as {ba}. Unlike the adaptation results, however, {spa} produced the same effect as {pa}: As a contrast stimulus, it acted in terms of its phonetic quality. Thus, there was a clear dissociation between adaptation and contrast effects.

On the basis of the Sawusch and Jusczyk (1981) findings, it is clear that contrast alone cannot account for selective adaptation effects; the major alternative to a fatigue account of adaptation has thus proved to be inadequate. Consequently, although there is, as yet, no direct evidence that adaptation arises from the fatigue of detectors, as originally proposed, this remains a viable alternative.

Before concluding this section, we should note that Diehl (1981), based on his findings of contrast effects, concluded that detectors are not a viable means of explaining the phenomena of speech perception, and from this further concluded that the signal does not undergo an initial analytic process. However, the rejection of a detector theory of speech perception does not necessarily entail the assumption that the signal is treated wholistically, since perceptual mechanisms other than fatigable detectors could perform an initial analysis. Particularly strong evidence for an initial analysis of the signal comes from a recent adaptation study by Miller (1981). Her test series included a {bae-pae} voicing series and a {bae-wae} manner series. The adapting stimuli included the endpoint {bae}, which was the same in both series, and the syllable {dae}. With respect to the {bae-pae} voicing series, since {d}, like {b}, is a voiced consonant, {dae} should behave as an adaptor like {bae}, shifting the boundary toward the {bae} (voiced) end of the series. Analogously, on the {bae-wae} series, since {d}, like {b}, is a consonant, {dae} like {bae} should produce a shift in category boundary toward the {bae} (stop) end of the series. Both adaptors produced a shift along the two series in the predicted direction. However, for the {bae-pae} voicing series, {dae} produced less adaptation than {bae}, whereas for the {bae-wae} manner series, {dae} and {bae} produced the same amount of adaptation. Thus, whether or not the processing system treated {bae} and {dae} as equivalent adaptors depended on the test dimension, a finding which models based on the wholistic processing of segments or syllables would have considerable difficulty in accommodating (cf. Diehl, 1981), but which can be readily explained by a model that assumes an initial analysis of the signal according to featural dimensions.

V. Auditory versus phonetic feature detectors.

Closely related to the issue of the source of the adaptation effect is whether the detectors, as revealed by the adaptation technique, are truly linguistic in nature, serving to extract phonetic features from the signal, or whether they are operating at an auditory level, analyzing not phonetic features, but acoustic properties (features). In essence, is there any evidence that these detectors, by being sensitive to phonetic information regardless of its acoustic form, can provide at least an initial answer to the problem of perceptual constancy? If the feature detectors operate at a truly phonetic level, as opposed to an auditory level of analysis, then they must exhibit at least two characteristics. First, they should respond invariantly to a given cue for a phonetic distinction, irrespective of any context conditioned changes in the acoustic characteristics of the cue. Second, inasmuch as phonetic features (e.g., voicing) are cued by a myriad of cues and not just a single cue (e.g., VOT, first-format transition characteristics), a phonetic feature detector must be sensitive to each of the multiple cues that specifies a given phonetic feature. Research has been addressed specifically to each of these issues and we will consider them in turn.

Contextual dependencies.

One way to assess whether a given detector operates in a context-dependent fashion is to systematically change the adapting stimulus, with respect to a test series, along a given dimension. This, in fact, was the tack taken in the Eimas and Corbit (1973) study, discussed earlier. Recall that perception was assessed on a labial series, {ba-pa}, as a function of adaptation with adaptors that had an alveolar place of articulation value, {da} or {ta}, and, likewise, tests were made on an alveolar series, {da-ta}, as a function of labial adaptors, {ba} and {pa}. If the detectors respond to the critical VOT information irrespective of context--in this case, irrespective of the place value of the segment--then equally large boundary shifts should occur after adaptation with the cross-series adaptors (e.g., {ba} on {da-ta}) as with adaptors drawn from the test series (e.g., {da} on {da-ta}). And, indeed, that is what happened. Thus, on the basis of this first study, it seemed as if the detectors for VOT were operating independently of the place of articulation feature value, cued by the second- and third-format transitions. However, in a host of subsequent studies, it has been shown that adaptation with an adapting stimulus that differs in a feature value from the test series often (though not always, cf. Miller, 1981) results in a diminished adaptation effect. For example, in two studies Miller (1981) found that changing the place value of the adaptor resulted in reduced effects, contrary to the Eimas and Corbit findings. Moreover, it has been shown that analysis of the place feature is dependent on the voicing value of the segment (e.g., Cooper and Blumstein, 1974; Miller, 1981) and that dependencies exist between the analysis of place of articulation and manner of articulation (e.g., Miller and Eimas, 1977; Miller, 1981). Taken together, these results indicate that the analysis of the acoustic information specifying a given feature of a segment may be dependent on the value of the other features of the segment, despite the fact that the acoustic information specifying the particular feature is left relatively unaltered by the change in context.[8]

Another type of context-dependency that has been demonstrated is that between the analysis of the featural information in a consonant and the quality of the following vowel. For example, if the vowel of the adapting stimulus and test series differ (e.g., adapt with {bi} and test on {ba-da}), a diminished adaptation effect is obtained, compared to that which would be obtained if the vowels were identical (e.g., adapt with {ba} and test on {ba-da}). This is true for a number of feature contrasts, including voicing (e.g., Cooper, 1974b; Miller and Eimas, 1976) and place of articulation (e.g., Ades, 1974b; Bailey, 1975; Miller and Eimas, 1976). A particularly elegant demonstration of this type of dependency was provided by Cooper (1974b), who employed a contingent adaptation procedure similar to that which has been successfully used in vision research (e.g., McCollough, 1965). In this study he assessed perception along two test series {ba-pa} and {bi-pi}, as a function of adaptation with an alternating series, {da}-{ti}. If the detectors operate in a vowel-independent manner, then adaptation with {d}-- a voiced -- and with {t} -- a voiceless -- should cancel each other and either produce no effect on both test series or a small net shift in the same direction on both series. However, if the detectors are sensitive to the vowel context, opposing adaptation effects should occur on the two test series: the {da} adaptor should cause a shift in the identification function toward the voiced value on the {ba-pa} series and, simultaneously, the {ti} adaptor should cause a shift toward the voiceless value on the {bi-ti} series. The results supported the latter set of predictions, thereby indicating that the analyzers for voicing (cued by VOT) were dependent on the following vowel. Similar results for the dependency of the place feature on the following vowel have also been obtained using this technique (Miller and Eimas, 1976; Sawusch and Pisoni, 1978). Interestingly, however, the processing of information for the place value of an initial consonant in a consonant-vowel-consonant syllable is not dependent on the place value of the final consonant (Miller and Eimas, 1976), indicating that there is a limit to the context-dependency exhibited by the detectors (cf. Miller, 1981).

A third type of context-dependency that has been studied is based on the fact that the acoustic form of a particular cue, for example the second- and third-format transitions that specify place of articulation, may change radically across different contexts. Consider, for example, the syllable patterns shown schematically in Figure 3, used in a study by Ades (1974b). The transitions for {b} in syllable-initial position ([bae]) are rising, whereas for {b} in syllable-final position ({aeb}) they are falling. Similarly, the transitions for syllable-initial {d} ({dae}) are falling, whereas those for syllable-final {d} ({aed}) are rising. If the detectors sensitive to the transition cue for place are phonetic in nature and

operate in a context-independent manner, then adaptation with {b}
should produce a shift toward {b} on a series that varies from {b} to
{d}, regardless of whether the consonantal information is in initial
or final position. For example, adaptation with {bae} should not only
produce a boundary shift on {bae-dae} toward {bae}, but also a shift
toward {aeb} on the {aeb-aed} series. However, if the detectors are
sensitive to the form of the transitions, rather than the phonetic
feature specified, then little adaptation on the cross-series is
expected.

Ades (1974b) tested these predictions by assessing the effects of
adaptation with {bae}, {aeb}, {dae}, and {aed} and both a {bae-dae}
and {aeb-aed} test series. Briefly, he obtained the expected effects
when the test and adapting stimuli both contained initial or final
consonants, but found no systematic effects when the consonant of the
test series and the consonant of the adapting stimulus were in
different syllable positions. Thus the detectors did not respond in an
invariant manner to the common phonetic feature, but rather were
sensitive to the form of the transitional information (see Pisoni and
Tash, 1975, and Wolf, 1978, for further discussion of these findings).
Related studies have shown that the magnitude of adaptation is
directly related to the amount of spectral overlap between test and
adapting stimuli, thereby providing additional evidence for a detector
system operating in terms of acoustic, not phonetic information (e.g.,
Bailey, 1975; Sawusch, 1977; Sawusch and Jusczyk, 1981). Finally, it
has been shown that the detectors are also sensitive to such
contextual parameters as fundamental frequency (e.g., Ades, 1976) and
intensity (e.g., Ganong, 1976).

We can conclude from these studies that the detectors are almost
always sensitive to contextual information: they do not typically
respond in an invariant fashion to a given cue across different
contexts. We should also note that this is true both when the actual
form of the acoustic information changes as a function of context, as
in the case of the transition cue for place of articulation in initial
versus final consonants (see Figure 3), and when the form of the cue
itself does not (at least radically) change with context, as in the
case of the VOT cue for voicing in labial versus alveolar consonants
(see Figure 4). In Figure 4 spectrograms of four synthetically-
produced syllables, {ba}, {da}, {pa}, and {ta} are shown. The voicing
distinction between {ba}, {da}, and {pa}, {ta} is cued by a change in
VOT from 0 msec to +45 msec, whereas the place distinction between

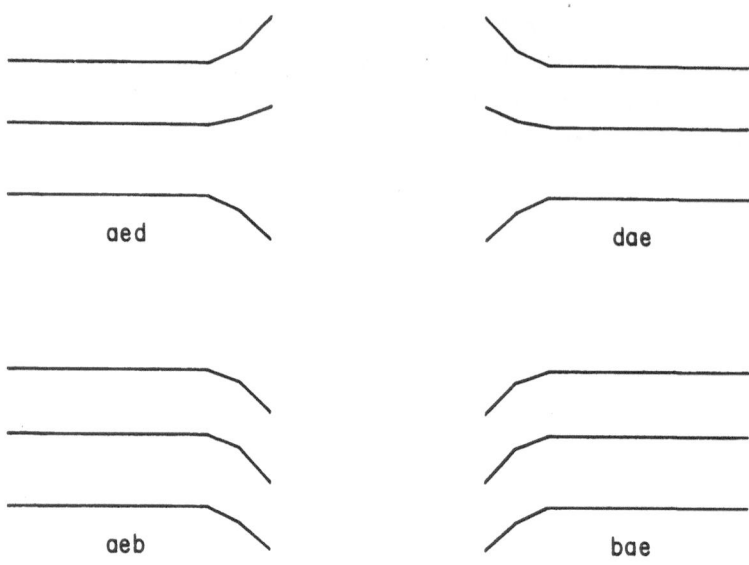

Figure 3: Schematic spectrograms of the extreme stimuli on the {aeb-aed} and {bae-dae} test continua. These stimuli were also used as the adapting sounds. (In a spectrographic display, frequency is shown as a function of time.) From Ades (1974b).

{ba}, {pa}, and {da}, {ta} is cued by a change in the starting frequency, direction, and extent of the second and third formant transitions. Inspection of the four syllables shows that the VOT cue itself does not change with a change in place, although, as we have seen, perceptual studies indicate that the mechanisms extracting VOT are sensitive to the place information.

A critical question is what form the detectors must take in order to handle context dependency, both when changes in context cause a radical change in the form of the cue and when they do not. Although the issue is far from resolved, we can offer some speculations. In the former case, one possibility is that different channels of analysis are responsible for the processing of the two different forms of the cue. In the latter case, one could postulate that the detectors are themselves multiply-tuned, sensitive both to the acoustic information specifying the feature under test (e.g., VOT),

[ba]   **[pa]**

[da]   **[ta]**

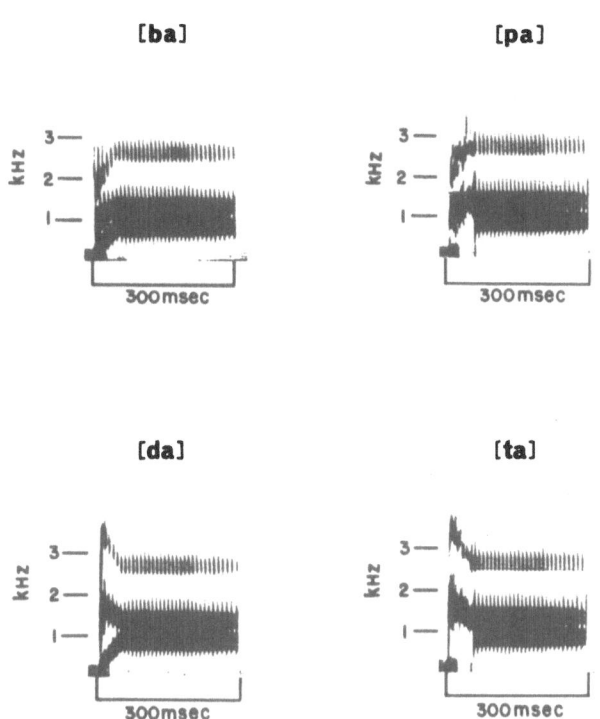

Figure 4: Spectrograms of four synthetically-produced syllables, {ba}, {pa}, {da}, and {ta}.

and the acoustic information specifying the context (e.g., the transitions cueing place of articulation).[9] In order to account for the perceptual equivalence that occurs with both types of contextual dependency, it must be further assumed that the outputs of the various channels of analysis for a particular feature are integrated at a higher level of processing.

Multiple cues.

From the discussion above it is evident that the first requirement of a feature detector system operating at a phonetic level, namely, that it respond invariantly to a given cue for a phonetic feature distinction regardless of context, has not been met. The detectors are sensitive to the acoustic context of the cue, which suggests that they are operating at a level before phonetic feature classification takes place, extracting not the phonetic feature value, per se, from the speech waveform, but rather the acoustic information relevant to that phonetic context, the detectors are capable of integrating the many, often diverse, cues that specify a given phonetic distinction. As will be shown, there is evidence that detectors may be sensitive to multiple cues and thus perform an integrative function required of phonetic detectors.

One of the first attempts to examine whether detectors are sensitive to more than one cue for phonetic feature was made by Cooper (1974c) in a study of voicing. As we have indicated, VOT is an important cue for the voicing feature in syllable-initial consonants. A second important cue for voicing is the presence or absence of a rapid spectral change after the onset of periodicity (Stevens and Klatt, 1974).[10] These two cues are typically confounded in natural speech and, in fact, in most experiments that vary VOT, including the Eimas and Corbit (1973) study. This occurs because the duration of the transitions is kept constant across stimuli in the series, and consequently as the VOT value increases, the amount of transition after the onset of voicing decreases. For example, consider a series, varying from [ba] to [pa], with initial transitions of 50 msec. As VOT increases from 0 msec to +50 msec, there is a concomitant decrease in the amount of voiced transition from 50 msec to 0 msec. At VOT values equal to or greater than +50 msec, there is no voiced transition. Cooper attempted to determine whether the analyzers, revealed by the selective adaptation procedure in the Eimas and Corbit study, were sensitive to absolute VOT, to duration of voiced transition, or some combination of both, the latter being evidence for an integrative detector.

The design of his study was as follows. The test stimuli were a series of syllables whose consonant varied in VOT from +5 msec ({ba}) to +55 msec ({pa}). Subjects were adapted on separate days with two adapting stimuli, both of which had VOT values of +25 msec. One had 40 msec of transition after voicing onset and was heard as a clear exemplar of a voiced consonant ({da}). The other had only 10 msec of voiced transition and was heard either as a very weak voiced stimulus or as intermediate between a voiced and voiceless stimulus ({da} or {ta}). If the detectors are sensitive to absolute VOT, then both adapting stimuli should have the same effect on the perception of the test series. However, if they are sensitive to voiced transition duration as well, they should have different effects; the syllable with a long voiced transition should shift the phonetic boundary toward the voiced end of the series, whereas the syllable with short voiced transitions should either shift the function only very slightly toward the voiced end of the series or have no effect. The adaptors had significantly different effects, indicating that the detector was not responsive to absolute VOT.

Two possible explanations were considered by Cooper. The first was that the detector is truly an integrative analyzer, integrating independent information about VOT, voiced transition duration and, most likely, other cues for voicing. This is the kind of integration we have argued is necessary for a detector operating in terms of a phonetic feature. The other alternative (and the one favored by Cooper) was that the detector system is not treating the cues as independent sources of information, but rather that the detector is responding to some weighted average of the cues, that is, to some higher-order invariant property defined jointly by the two cues. An example would be the ratio of voiceless (aspirated) transition to voiced transition (see Stevens and Klatt, 1974). The critical difference between the two explanations is whether the invariance is provided by the mechanism (in the case of an integrative analyzer) or is inherent in the signal (a higher-order invariant cue for voicing) and the detector is simply responsive to this invariant—a distinction we raised earlier in the paper.

Attempts have also been made to determine if the detectors subserving the analysis of place of articulation operate as integrative analyzers. One approach to this problem has been to use test and adapting stimuli that have the identical phonetic feature value, but different acoustic properties cueing that feature. For example, it has traditionally been assumed that there are two primary, independent cues for place of articulation in syllable-initial consonants, the initial burst and the initial formant transitions (e.g., Dorman, Studdert-Kennedy, and Raphael, 1977; Liberman et al., 1967). Diehl (1975) attempted to test the notion of an integrative analyzer by assessing perception along a series varying in place, in which place was cued solely by transitions, as a function of adaptation with stimuli for which place was cued by bursts. Inasmuch as the burst and transition cues presumably had no common acoustic information, any adaptation effect obtained would be evidence for a detector sensitive to both cues. Diehl found significant adaptation by the burst-cued stimuli and, in a more recent and complete study along the same lines, Ganong (1978) has provided additional evidence for detectors responsive to both bursts and transitions.

However, as in the case of voicing, we are left with two possible explanations of these results. The first is that the burst and transitions are indeed independent cues and that the detectors are sensitive to and integrate both acoustic features. The other, which is also consistent with the existing data, is that the transitions and burst are not independent cues, but rather contribute to a higher-order property of the signal that specifies place of articulation. In a series of papers, Stevens and Blumstein (1978, 1981; Blumstein and Stevens, 1979, 1980) have argued that the burst and transitions contribute to a global acoustic property--the shape of the spectrum at stimulus onset--and that it is the onset spectrum that specifies place. Evidence consistent with this explanation was recently obtained by Stevens and Blumstein in a study involving selective adaptation (Blumstein, Stevens, and Nigro, 1977). Thus for place, as for voicing, it is still unclear whether the detectors tapped by selective adaptation are truly integrative detectors, integrating diverse and separate acoustic cues for a phonetic distinction, or whether they analyze higher-order information in the signal that specifies the phonetic feature value.

VI. Concluding remarks

Although the application of a feature detector model to speech and the empirical testing of this model have a relatively short history, this research has provided considerable information both about the adequacy and the limits of such a model for speech perception. First, the numerous studies using the selective adaptation technique provide, we believe, strong evidence for the existence of channels of analysis that are sensitive to information that is critical for the specification of phonetic features.

Second, from the research on the integration of cues for a single phonetic feature within a given context, we can conclude that these mechanisms are most likely not sensitive only to "simple" properties of the speech wave-form, such as particular formant transitions, but that they are sensitive to an entire complex of cues that signals a phonetic feature value in a given acoustic context. To the extent that the detectors do operate in this fashion, they may account for one aspect of perceptual constancy, namely, how seemingly diverse and multiple cues for a phonetic feature yield equivalent perceived phonetic segments. As we have discussed earlier, the manner in which this is accomplished is not clear. One possibility is that the detectors do integrate the many independent sources of acoustic information that cue a particular phonetic feature. Another, however, is that they are sensitive to higher-order "global" properties that themselves form the basis of the phonetic decision, comparable to the mechanisms tuned to the complex species-specific vocalizations found in some nonhuman organisms (e.g., Wollberg and Newman, 1972; and see Newman, 1977, for a review of these findings). If the first possibility is true, then, at least within a single context (and it is important to emphasize that there are contextual restrictions), the detectors are providing the invariance; if the second is true, then the invariance is inherent in the signal. The detector model itself does not favor one approach over the other to the solution of the problem of perceptual constancy, but is compatible with either. Research that focuses both on the possible higher-order properties that could serve as invariant cues within a single context, and on the possible integrative operation of the detectors, can potentially help choose between these two alternatives.

Finally, from the studies on context-dependency, it is apparent that the channels of analysis for speech are extremely sensitive to context: They do not respond invariantly to a given cue across contexts, either when the form of the cue changes with context or when it remains fairly stable. We must conclude, therefore, that these mechanisms are tuned to acoustic, not phonetic, information. And, since they are not true "phonetic feature detectors," they do not account for perceptual constancy across contexts, leaving one of the major puzzles of speech perception unsolved. What the detectors do appear to do is extract from the signal the relevant acoustic information that is used by higher levels of processing to assign phonetic feature values. The nature of these levels of processing-- mechanisms that can assign phonetic feature values on the basis of context-conditioned information and in so doing provide the basis for perceptual constancy--remains a challenging problem for future research.

Footnotes

[1]This paper was originally prepared for presentation at the Symposium on the Role of Feature Detectors in the Recognition of Pattern and Form, held in Austin, Texas, in 1979, and was revised for publication in 1981. Preparation of the paper and the authors' research reported herein were supported by NIH grants NS 14394 to J.L. Miller and HD 05331 to P.D. Eimas.

[2]Although there is currently much debate over whether a nonhuman species, such as the chimpanzee or gorilla, is capable of learning a human or human-like communication system (e.g., Garner and Gardner, 1971; Premack, 1971; Terrace, Petito, Sanders, and Bever, 1979; cf. Harnad, Steklis, and Lancaster, 1976), it is clearly the case that none has yet acquired a full-fledged linguistic system. Moreover, the course of acquisition evidenced by the nonhuman organisms does not resemble that shown by the human child, who quite rapidly and with little apparent effort acquires the language to which he or she is exposed without the benefit of direct instruction.

[3]There is a history of controversy over whether the phonetic segments themselves are recovered during the perceptual process or whether other units, such as the syllable, comprise the basic units of speech perception. For a discussion of this issue and, in particular, arguments supporting the claim that phonetic segments are functional units in perception, see Studdert-Kennedy, 1976.

[4]We do not mean to imply that during normal conversation all decisions regarding which phonetic segments are heard are based solely on the acoustic information in the speech signal. Phonetic decisions are also influenced by information from higher levels of language-- phonological, syntactic, semantic, and pragmatic information; that is, the final decision is arrived at by both bottom-up and top-bottom processing (e.g., Ganong, 1980; Marslen-Wilson and Welsh, 1978). However, the fact that higher level information influences phonetic perception does not vitiate the assumption that an early stage of speech processing involves the extraction of features by detector mechanisms.

[5]We have used the terms feature detector, property detector, and channel of analysis more or less interchangeably in this chapter.

[6]The actual VOT value at the voiced-voiceless boundary is not constant, but depends on such contextual factors as the place of articulation of the consonant (e.g., Miller, 1977a), the following vowel (e.g., Cooper, 1974b), and the duration of the syllable, presumably cueing rate of articulation (e.g., Summerfield, 1975).

[7]Manner of articulation is an additional feature that distinguishes consonants in traditional classification systems. There are actually a number of manner distinctions including, for example, the distinction between nasal and consonants (e.g., {m} versus {b}) and that between consonants and semivowels (e.g., {b} versus {w}).

[8]Experiments using paradigms other than selective adaptation have also revealed dependencies in the analysis of features. For example, as described in footnote 5, the VOT value at the voiced-voiceless boundary depends, among other things, on the place of articulation of the consonant, and, conversely, the place boundary depends on the voicing of the consonant (Miller, 1977a). Dependencies of this nature present a serious problem for any perceptual system; that is, it is necessary to explain how a given acoustic value is assigned a different phonetic value as a function of the context in which it resides.

[9]This would of course require sets or populations of detectors, each of which, while tuned to a specific feature (e.g., VOT), would have slightly different specificities with respect to the information which defines the context (e.g., place and manner information, speaker, and the like). It is of interest to note that the channels of analysis for visual information, like those for speech, also appear to be highly context-dependent and tuned to multiple aspects of the stimulus (see Eimas and Miller, 1978) and that systems of analysis based on populations of neurons have been put forth to accommodate problems arising from altered responsivity in various contexts and form the existence of broadly tuned channels (cf., Erickson, 1974).

[10]Investigations subsequent to that of Cooper on the spectral onset cue for voicing have shown that the effective cue may not be the presence or absence of rapid spectral change after voicing onset, _per se_, but the frequency value of the first formant transition at voicing onset (Lisker, 1975; Summerfield and Haggard, 1977).

References

Abbs, J.H., and Sussman, H.M. 1971. Neurophysiological feature detectors and speech perception: A discussion of theoretical implications. Journal of Speech and Hearing Research, 14, 23-36.

Abramson, A.S, and Lisker, L. 1970. Discriminability along the voicing continuum: Cross-language tests. In Proceedings of the Sixth International Congress of Phonetic Sciences, Prague, 1967. Prague: Academia.

Ades, A.E. 1974. Bilateral component in speech perception? Journal of the Acoustical Society of America, 56, 610-617. (a)

Ades, A.E. 1974. How phonetic is selective adaptation? Experiments on syllable position and vowel environment. Perception and Psychophysics, 16, 61-67. (b)

Ades, A.E. 1976. Adapting property detectors for speech perception. In R. Wales and E. Walker (Eds.), New Approaches to Language Mechanisms. Amsterdam: North Holland.

Ainsworth, W.A. 1977. Mechanisms of selective feature adaptation. Perception and Psychophysics, 21, 365-370.

Anderson, F. 1975. Some implications for the operation of feature detectors in speech perception: Use of identification response time as a converging operation. Unpublished Ph.D. thesis, Brown University.

Anstis, S.M. 1975. What does visual perception tell us about visual coding? In M.S. Gazzaniga and C. Blakemore (Eds.), Handbook of Psychobiology, New York: Academic Press.

Bailey, P.J. 1975. Perceptual adaptation in speech: Some properties of detectors for acoustical cues to phonetic distinctions. Unpublished Ph.D. thesis, University of Cambridge.

Blakemore, C. 1975. Central visual processing. In M.S. Gazzaniga and C. Blakemore (Eds.), Handbook of Psychobiology, New York: Academic Press.

Blakemore, C., and Campbell, F.W. 1969. On the existence of neurones in the human visual system selectively sensitive to the orientation and size of retinal images. Journal of Physiology, 203, 237-260.

Blumstein, S.E., and Stevens, K.N. 1979. Acoustic invariance in speech production: Evidence from measurements of the spectral characteristics of consonants. Journal of the Acoustical Society of America, 66, 1007-1017.

Blumstein, S.E., and Stevens, K.N. 1980. Perceptual invariance and onset spectra for consonants in different vowel environments. Journal of the Acoustical Society of America, 67, 648-662.

Blumstein, S.E., Stevens, K.N., and Nigro, G.N. 1977. property detectors for bursts and transitions in speech perception. Journal of the Acoustical Society of America, 61, 1301-1313.

Chomsky, N. 1972. Language and Mind. New York: Harcourt, Brace, Jovanovich.

Chomsky, N. 1975. Reflections on Language. New York: Random House.

Chomsky, N., and Halle,M. 1968. The Sound Pattern of English. New York: Harper and Row.

Cole, R.A., and Scott, B. 1974. Toward a theory of speech perception. Psychological Review, 4, 348-374.

Cooper, W.E. 1974. Adaptation of phonetic feature analyzers for place of articulation. Journal of the Acoustical Society of America, 56, 617-627. (a)

Cooper, W.E. 1974. Contingent feature analysis in speech perception. Perception and Psychophysics, 16, 201-204. (b)

Cooper, W.E. 1974. Selective adaptation for acoustic cues of voicing in initial s. Journal of Phonetics, 2, 303-313.

Cooper, W.E. 1975. Selective adaptation to speech. In f. Restle, R.M. Shiffrin, N.J. Castellan, H. Landman, and D.B. Pisoni (Eds.), Cognitive Theory, 1, Potomac, Maryland: Erlbaum Assoc..

Cooper, W.E. and Blumstein, S.E. 1974. A labial feature analyzer in speech perception. Perception and Psychophysics, 15, 591-600.

Cooper, W.E., Ebert, R.R., and Cole, R.A. 1976. Perceptual analysis of consonants and glides. Journal of Experimental Psychology: Human Perception and Performance, 2, 92-104.

Darwin, C.J. 1976. The perception of speech. In E.C. Carterette and M.P. Friedman (Eds.), Hankbook of Perception. New York: Academic Press.

Diehl, R.L. 1975. The effect of selective adaptation on the identification of speech sounds. Perception and Psychophysics, 17, 48-52.

Diehl, R.L. 1976. Feature analyzers for the phonetic dimension vs. continuant. Perception and Psychophysics, 19, 267-272.

Diehl, R.L., 1981. Feature detectors for speech: A critical reappraisal. Psychological Bulletin, 89, 1-18.

Diehl, R.L., Elman, J.L., and McCusker, S.B. 1978. Contrast effects on consonant identification. Journal of Experimental Psychology: Human Perception and Performance, 4, 599-609.

Diehl, R.L., Lang, M., and Parker, E.M. 1980. A further parallel between selective adaptation and contrast. Journal of Experimental Psychology: Human Perception and Performance, 6, 24-44.

Dorman, M., Studdert-Kennedy, M., and Raphael, L.J. 1977. Consonant recognition: Release bursts and formant transitions as functionally equivalent, context-dependent cues. Perception and Psychophysics, 22, 109-122.

Eimas, P.D., and Corbit, J.D. 1973. Selective adaptation of linguistic feature detectors. Cognitive Psychology, 4, 99-109.

Eimas, P.D., Siqueland, E.R., Jusczyk, P., and Vigorito, J. 1971. Speech perception in infants. Science, 171, 303-306.

Eimas, P.D., Tartter, V.C., Miller, J.L., and Keuthen, N. 1978. Asymmetric dependencies in processing phonetic features. Perception and Psychophysics, 23, 12-20.

Elman, J.L. 1979. Perceptual origins of the phoneme boundary effect and selective adaptation to speech: A signal detection analysis. Journal of the Acoustical Society of America, 65, 179-189.

Erickson, R.P. 1974. Parallel "population" neural coding in feature extraction. In F.O. Schmitt and F.G. Worden (Eds.), The Neurosciences: Third Study Program. Cambridge, Mass.: MIT Press.

Fant, G. 1960. Acoustic Theory of Speech Production. The Hague: Mouton.

Foreit, K.G. 1977. Linguistic relativism and selective adaptation for speech: A comparative study of English and Thai. Perception and Psychophysics, 21, 347-351.

Ganong, W.F.III 1976. Amplitude contingent selective adaptation to speech. Journal of the Acoustical Society of America, 59, S26. (A)

Ganong, W.F.III 1978. The selective adaptation effects of burst-cued s. Perception and Psychophysics, 24, 71-83.

Ganong, W.F.III 1980. Phonetic categorization in auditory work perception. Journal of Experimental Psychology: Human Perception and Performance, 6, 110-125.

Gardner, B.T., and Gardner, R.A. 1971. Two-way communication with an infant chimpanzee. In A. Schrier and F. Stollnitz (Eds.), Behavior of Nonhuman Primates. New York: Academic Press.

Gibson, J.J. 1979. The Ecological Approach to Visual Perception. Boston: Houghton Mifflin Co..

Gibson, J.J., Osser, H., Schiff, W., and Smith, J. 1963. An analysis of critical features of letters, tested by a confusion matrix. In Final Report on a Basic Research Program on Reading. Cooperative Research Project No. 639. Cornell University and U.S. Office of Education.

Greenberg, J.H., and Jenkins, J.J. 1964. Studies in the psychological correlates of the sound system of American English. Word, 22, 157-177.

Haggard, M.P. 1970. The use of voicing information. Speech Synthesis and Perception, 2, Cambridge: Psychological Laboratory, Cambridge University.

Harnad, S., Steklis, H., and Lancaster, J. 1976. Origins and Evolution of Language and Speech. Annals of the New York Academy of Sciences, Vol. 280, New York: New York Academy of Sciences.

Helson, H. 1964. Adaptation-level theory. New York: Harper and Row.

Helson, H. 1971. Adaptation-level theory: 1970- and after. In M.H. Appley (Ed.). Adaptation-level theory. New York: Academic Press.

Hoy, R.R. 1974. Genetic control of acoustic behavior in crickets. American Zoologist, 14, 1067-1080.

Hoy, R.R., Hahn, J., and Paul, R.C. 1977. Hybrid cricket auditory behavior: Evidence for genetic coupling in animal communication. Science, 195, 82-48.

Hubel, D.H., and Wiesel, T.N. 1962. Receptive fields, binocular interaction and functional architecture in the cat's visual cortex. Journal of Physiology, 160, 106-154.

Hubel, D.H., and Wiesel, T.N. 1965. Receptive fields and functional architecture in two non-striate visual areas (18 and 19) of the cat. Journal of Neurophysiology, 28, 229-289.

Hubel, D.H., and Wiesel, T.N. 1968. Receptive fields and functional architecture of monkey striate cortex. Journal of Physiology, 195, 215-243.

Liberman, A.M. 1970. The grammars of speech and language. Cognitive Psychology, 1, 301-323.

Liberman, A.M., Cooper, F.S., Harris, K.S., and MacNeilage, P.F. 1963. A motor theory of speech perception. In G. Fant (Ed.), Proceedings of the Speech Communication Seminar. Stockholm: Royal Institute of Technology. Speech Transmission Laboratory.

Liberman, A.M., Cooper, F.S., Shankweiler, D.P., and Studdert-Kennedy, M. 1967. Perception of the speech code. Psychological Review, 74, 431-461.

Lieberman, P. 1970. Towards a unified phonetic theory. Linguistics Inquiry,1, 307-322.

Lisker, L. 1975. Is it VOT or a first-formant transition detector? Journal of the Acoustical Society of America, 57, 1547-1551.

Lisker, L. and Abramson, A.S. 1964. A cross-language study of voicing of initial s: Acoustical measurements. Word, 20, 384-422.

Marslen-Wilson, W.D., and Welsh, A. 1978. Processing interactions and lexical access during work-recognition in continuous speech. Cognitive Psychology, 10, 29-63.

Mattingly, I. April 1968. Synthesis by rule of general American English. In Supplement to Status Report on Speech Research. New Haven, Conn.: Haskins Laboratories.

McCollough, C. 1965. Color adaptation of edge-detectors in the human visual system. Science, 149, 1115-1116.

Miller, G.A., and Nicely, P. 1955. An analysis of perceptual confusion among some English consonants. Journal of the Acoustical Society of America, 27, 338-352.

Miller, J.L. 1975. Properties of feature detectors for speech: Evidence from the effects of selective adaptation on dichotic listening. Perception and Psychophysics, 18, 389-397.

Miller J.L. 1975. Nonindependence of feature processing in initial consonants. Journal of Speech and Hearing Research, 20, 519-528. (a)

Miller, J.L. 1977. Properties of feature detectors for VOT: The voiceless channel of analysis. Journal of the Acoustical Society of America, 62, 641-648. (b)

Miller, J.L. 1981. Phonetic perception: Evidence for context-dependent and context-independent processing. Journal of the Acoustical Society of America, 69, 822-831.

Miller, J.L., and Eimas, P.D. 1976. Studies on the selective tuning of feature detectors for speech. Journal of Phonetics, 4, 119-127.

Miller, J.L., and Eimas, P.D. 1977. Studies on the perception of place and manner of articulation: A comparison of the labial-alveolar and nasal- distinctions. Journal of the Acoustical Society of America, 61, 835-845.

Moffitt, A.R. 1971. Consonant cue perception by twenty to twenty-four week old infants. Child Development, 42, 717-713.

Morse, P.A. 1972. The discrimination of speech and nonspeech stimuli in early infancy. Journal of Experimental Child Psychology, 14, 477-492.

Neisser, U. 1967. Cognitive Psychology. New York: Appleton-Century-
    Crofts.

Newman, J.D. 1977. Biological filtering and neural mechanisms. In T.H.
    Bullock (Ed.), Recognition of Complex Acoustic Signals: Report
    of the Dahlem Workshop on Recognition of Complex Acoustic
    Signals. Berlin: Dahlem Konferenzen.

Parker, F. 1977. Distinctive features and acoustic cues. Journal of
    the Acoustical Society of America, 62, 1051-1054.

Pisoni, D.B. 1978. Speech perception. In W.K. Estes (Ed.), Handbook of
    Learning and Cognitive Processing, VI, Hillsdale, New Jersey:
    Lawrence Erlbaum Assoc..

Pisoni, D.B., and Tash, J. 1975. Auditory property detectors and
    processing place features in consonants. Perception and
    Psychophysics, 18, 401-408.

Premack, D. 1971. On the assessment of language competence in
    chimpanzees. In A.Am Schrier and F. Stollnitz (Eds.), Cognitive
    Processes of Nonhuman Primates. New York: Academic Press.

Remez, R.E. 1979. Adaptation of the category boundary between speech
    and nonspeech: A case against feature detectors. Cognitive
    Psychology, 11, 38-57.

Sawusch, J.R. 1977. Peripheral and central processes in selective
    adaptation of place of articulation in consonants. Journal of
    the Acoustical Society of America, 62, 738-750.

Sawusch, J.R., and Jusczyk, P. 1981. Adaptation and contrast in the
    perception of voicing. Journal of Experimental Psychology: Human
    Perception and Performance, 7, 408-421.

Sawusch, J.R., and Pisoni, D.B. 1978. Simple and contingent adaptation
    effects for place of articulation in consonants. Perception and
    Psychophysics, 23, 125-131.

Sawusch, J.R., Pisoni, D.B., and Cutting, J.E. 1974. Category
    boundaries for linguistic and nonlinguistic dimensions of the
    same stimuli. Journal of the Acoustical Society of America, 55,
    S55. (A)

Simon, H.J., and Studdert-Kennedy, M. 1978. Selective anchoring and
    adaptation of phonetic and nonphonetic continua. Journal of the
    Acoustical Society of America, 64, 1338-1357.

Smith, P.T. 1973. Feature-testing models and their application to
    perception and memory for speech. Quarterly Journal of
    Experimental Psychology, 25, 511-534.

Stevens, K.N. 1960. Toward a model for speech recognition. Journal of
    the Acoustical Society of America, 32, 47-55.

Stevens, K.N. 1975. The potential role of property detectors in the perception of consonants. In G. Fant and M.A.A. Tatham (Eds.), _Auditory Analysis and Perception of Speech_. London: Academic Press.

Stevens, K.N., and Blumstein, S.E. 1978. Invariant cues for place of articulation in consonants. _Journal of the Acoustical Society of America_, 64, 1358-1368.

Stevens, K.N., and Blumstein, S.E. 1981. The search for invariant acoustic correlates of phonetic features. In P.D. Eimas and J.L. Miller (Eds.), _Perspectives on the Study of Speech_. Hillsdale: Erlbaum Assoc..

Stevens, K.N., and House, A.S. 1972. Speech Perception. In j. Tobias (Ed.), _Foundations of Modern Auditory Theory_, Vol. II. New York: Academic Press.

Stevens, K.N., and Klatt, D.H. 1974. Role of formant transition in the voiced-voiceless distinction for s. _Journal of the Acoustical Society of America_, 55, 653-659.

Stromeyer, C.F. III 1978. Form-color aftereffects in human vision. In R. Held, H. Leibowitz, and H.L. Teuber (Eds.), _Handbook of Sensory Physiology: Perception_. Berlin: Springer-Verlag.

Studdert-Kennedy, M. 1974. The perception of speech. In T.A. Sebeok (Ed.), _Current Trends in Linguistics_. The Hague: Mouton.

Studdert-Kennedy, M. 1976. Speech Perception. In N.J. Lass (Ed.), _Contemporary Issues in Experimental Phonetics_. New York: Academic Press.

Studdert-Kennedy, M., and Shankweiler, D.P. 1970. Hemispheric specialization for speech perception. _Journal of the Acoustical Society of America_, 48, 570-594.

Summerfield, A.Q. 1975. Information processing analyses of perceptual adjustments to source and context variables in speech. Unpublished Ph.D. thesis, Queen's University of Belfast.

Summerfield, Q., and Haggard, M. 1977. On the dissociation of spectral and temporal cues to the voicing distinction in initial consonants. _Journal of the Acoustical Society of America_, 62, 435-448.

Terrace, H.S., Petito, L.A., Sanders, R.J., and Bever, T.G. 1979. Can an ape create a sentence? _Science_, 206, 891-902.

Trehub, S.A., and Rabinovitch, M.S. 1972. Auditory-linguistic sensitivity in early infancy. _Developmental Psychology_, 6, 74-77.

Ward, W.D. 1973. Adaptation and fatigue. In J. Jerger (Ed.), _Modern Developments in Audiology_. New York: Academic Press.

Werner, G. 1974. Neural information processing with stimulus feature extractors. In F.O. Schmitt and F.G. Worden (Eds.), The Neurosciences Third Study Program. Cambridge: MIT press.

Wolf, C. 1978. Perceptual invariance for consonants in different positions. Perception and Psychophysics, 24, 315-326.

Wollberb, A., and Newman, J.D. 1972. Auditory cortex of squirrel monkey: Response patterns of single cells to species-specific vocalizations. Science, 175, 212-214.

THE PHYSICS OF VISUAL PERCEPTION

F.W. Campbell and Mark Lawden

Physiological Laboratory
Cambridge CB2 3EG
England

Any temporal or spatial stimulus can be characterised by its Fourier transform (Taylor, 1965). By varying the contrast and spatial frequency (fineness) of a grating, kept at constant mean luminance, it is possible to define quantitatively those parts of a visual scene that are available to our eye/brain. This technique, and the results obtained in man and in the cat are described by Campbell and Maffei (1974). The results are summarized in Figure 1. The shaded area is the invisible world where ectoplasm, fairies and ghosts haunt each other in privacy. How does this approach help us to understand the visual world that we can see (the unshaded area)? Why are we not aware of the undetecable portions of the scene? These problems troubled Helmholtz and Mach (Ratliff, 1965).

Envisage examining a standard optometrist's eye test chart. Lines that are too small for us to resolve still seem to be composed of high contrast black letters. Thus, there seems to be some mechanism to ensure contrast constancy, so that as we move about the world, the contrast of objects does not change (Georgeson and Sullivan, 1975).

Campbell and Robson (1968) showed that the contrast threshold for gratings with a sinusoidal profile was different from those with a square-wave profile. These observations were extended into the suprathreshold domain by Campbell, Howell and Johnstone (1978). For gratings with spatial frequency greater than 1 c/d, the square-wave grating is perceived slightly better by a factor of $4/\pi$. If we accept that in this spatial frequency range the visual system performs a quasi-Fourier analysis on the image, this is precisely the expected result, for this is the power of the fundamental in a square-wave. Indeed, it is difficult to see what other mechanism could account for this finding.

For frequencies less than 1 c/d, the visual system behaves quite differently. At any of these low frequencies, square waves of all

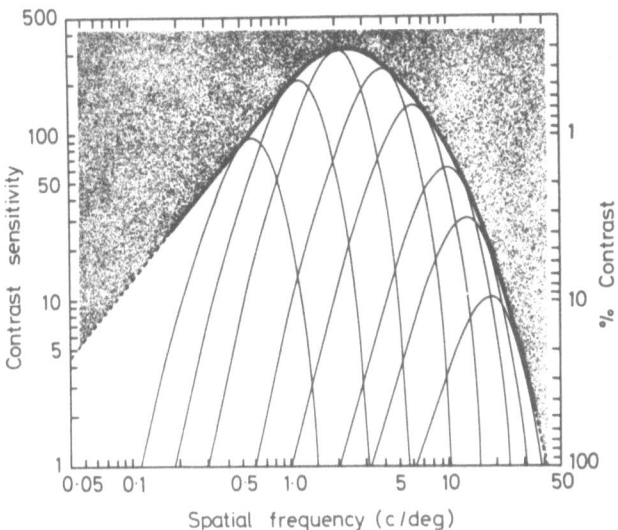

Figure 1: The thick curve represents the contrast sensitivity (defined as reciprocal threshold contrast) of the human visual system to a sinusoidal grating, plotting against spatial frequency. The shaded area must always remain invisible to us unless the spatial frequency content of the image is shifted into the visible domain by optical means, such as the microscope. The lighter curves represent channels sensitive to a narrow range of spatial frequencies.

spatial frequencies have a constant low (0.25%) contrast threshold. Even at a spatial frequency of 1 cycle per 180 degrees (i.e., a single edge) this remains valid. As the spatial frequency is clearly irrelevant in this instance, it seems that the square-wave is detected by virtue of its individual edges rather than as a gestalt grating. Sinusoidal gratings of these low frequencies are greatly attenuated or invisible. Indeed, they can be removed from a square-wave grating, altering its luminance profile dramatically, but without affecting its appearance or detectability. This effect is analagous to the missing fundamental phenomenon in audition (Goldstein, 1973), and has a neurophysiological substrate (Maffei, L., Morrone, C., Pirchio, M. and Sandini, G., 1979).

A sinusoidal grating is physically infinitely blurred, while a square-wave is infinitely sharp. A sine-wave has only a single fundamental frequency in its Fourier spectrum containing all the wave's power. A square wave has, in addition, a shower of higher odd harmonics of linearly decreasing amplitude. Campbell, Howell and Johnstone (1978) have shown that low frequency square gratings are perceived as true square-waves, providing that they possess the first few higher harmonics (in appropriate phase). The auditory analogue of this is that the click of a metronome is detected by the neurones tuned to high temporal frequencies. Similar neurones tuned to a band of spatial frequencies, approximately one octave wide, exist in the visual cortex of the cat (review by Maffei, 1978) and the elegant work of De Valois presented at this meeting has shown that the monkey visual cortex is similarly organised. Indeed, the important discovery by Hubel and Wiesel (1962) that cells in the visual cortex are sensitive only to a narrow range of stimulus orientations is immediately suggestive of a Fourier analytical process. The existence of channels sensitive to particular bands on spatial frequencies of particular orientation can be demonstrated psychophysically in man by adaptation experiments (see review by Braddick, Campbell and Atkinson, 1978). Indeed the first evidence for channels was found psychophysically by Campbell and Robson (1968) who stated that "a picture emerges of functionally separate mechanisms in the visual nervous system each responding maximally at some particular spatial frequency and hardly at all at spatial frequencies differing by a factor of two. The frequency selectivity of these mechanisms must be determined by integrative processes in the nervous system and they appear to a first approximation at least, to operate linearly."

Figure 2 illustrates a picture of a tank that has been analysed by three broad-band channels, respectively sensitive to the low, medium and high-frequency components of the original image. An enemy soldier would be most interested in the low-frequency components and having established that a tank is approaching him, will turn his attention to survival in the undergrowth. The tank troop commander, however, will be most interested in the intermediate frequency components, which reveal the tank type and number, whilst the sergeant of the maintenance wing will examine the high-frequency components for signs of damage to the trackwork.

The channel model of the visual system raises many interesting questions. For example, Pirenne (1967) has raised an important question when he writes, "It is a familiar fact that on a moonless

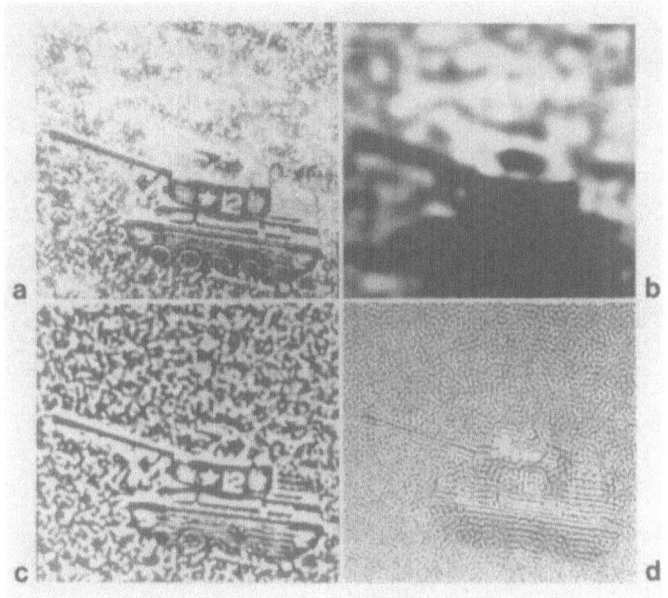

Figure 2: (a) Original photographic image. (b) Same image filtered to pass low spatial frequencies only. (c) As (b) but with only medium spatial frequencies passed. (d) as (b) but with only high spatial frequencies passed. (Courtesy of H.C. Andrews).

night, our vision not only is very blurred, but is also colourless". If a myope removes his spectacles, high spatial frequencies suffer greater attenuation than do low spatial frequencies. In this case, the visual scene does indeed look very blurred. As the illumination of a scene is reduced, high spatial frequencies are again attenuated much more than low, due to the shortage of photons available to the visual system, over the eye's integration time (0.1 second), and we might expect Pirenne's prediction of blurr to be verified. However, our own observations have led us to the conclusion that at night the visual scene, though devoid of fine detail, is quite definitely not blurred.

A low frequency sine-wave grating (less than 1 c/d) is always discriminably different from a square-wave grating, even near threshold. This holds to the lowest luminance level. This must mean that some of the higher harmonics of the square-wave are being detected. It can also be shown readily that if the eye is defocussed by as little as + O.5D, the resulting optical blurr can be detected (night myopia).

Due to photon shortage at these dim illuminations, the neurones responding to high frequencies which can function under photopic conditions, fail to operate, yet we do not notice their absence. The visual system must be aware that these neurones can never fire at low levels, and assumes, if the question arises (as in the perception of an edge or square-wave), that the undetecable frequencies are present. Campbell, et al (1978) concluded that "the visual input appears to be analysed into its Fourier components and if these constitute a square sequence with no above-threshold components missing, then the perception of a square-wave results. This solves the problem, considered by Helmholtz and Mach (Ratliff, 1965, p. 265), of why the aberrations of the emmetropic eye do not result in all edges appearing blurred. Only when the blurring is great enough to remove independently detectable high frequency components does the blurr become detectable." Likewise, under scotopic conditions, a square-wave grating always appears to be square because any undetectable missing higher harmonics are not noticed as absent.

Another feature of a moonless night is that, at least in the clear air of Texas, the sky is full of stars. It is our opinion that all stars, whether seen in the fovea or the periphery, look like point sources. This raises an interesting question, for the acuity of vision falls off markedly as eccentricity from the fovea increases. One might expect that when regarding the "inverted bowl they call the sky" (Fitzgerald, 1859) we would see the stars as point sources in the fovea, but as broader and broader smudges as their images fall upon peripheral retinal locations of lower and lower acuity. The fact that we do not suggests that the visual system perceives a small stimulus as a point source, until that stimulus reaches a size sufficient that its true size shall be signalled by the available channel mechanisms. Thereafter, of course, the true size is perceived, and the objects retain the same apparent size, wherever upon the retina their image falls. The channel system is organised such that where its limited sensitivity intudes, the stimulus is perceived in its most probable form, be that as a sharp edge or square-wave, or as a star.

151

References

Andrews, H.C. (1972). Digital computers and image processing. Endeavour, 31, 88-94.

Braddick, O., Campbell, F.W. and Atkinson, J. (1978) Channels in Vision: Basic aspects. Handbook of Sensory Physiology. Vol VIII. Chapter 1, pp. 3-38, Springer-Verlag.

Campbell, F.W., Howell, E.R. and Johnstone, J.R. (1978). A comparison of threshold and suprathreshold appearance of gratings with components in the low and high spatial frequency range. J. Physiol., 284, 193-201.

Campbell, F.W. and Maffei, L. (1974). Contrast and Spatial Frequency. Sci. Am., 231, 106-115.

Campbell, F.W. and Robson, J.G. (1968). Application of Fourier analysis to the visibility of gratings. J. Physiol., 197, 551-566.

Fitzgerald, E., Rubaiyat of Omar Khayyam (1859).

Georgeson, M.A. and Sullivan, G.D. (1975). Contrast Constancy: deblurring in human vision by spatial frequency channels. J. Physiol., 252, 627-656.

Goldstein, J.L. (1973). An optimum process theory for the central information of the pitch of complex tones. J. Acoust. Soc. Am., 54, 1496-1516.

Hubel, D.H. and Wiesel, T.N. (1962). Receptive fields, binocular interaction and functional architecture in the cat's visual cortex. J. Physiol., 160, 106-154.

Maffei, L. (1978). Spatial frequency channels: Neural mechanisms. Handbook of Sensory Physiology. Vol VIII. Chapter 1, pp. 39-66, Springer-Verlag.

Maffei, L., Morrone, C., Pirchio, M. and Sandini, G. (1979). Responses of visual cortical cells to periodic and non-periodic stimuli. J. Physiol., (in press).

Pirenne, M. (1967). Vision and the Eye. P. 36, Chapman and Hall.

Ratliff, F. (1965). Mach Bands: Quantitative Studies on Neural Networks in the Retina. San Francisco: Holden-Day.

Taylor, C.A. (1965). The Physics of Musical Sounds. The English Universities Press.

EARLY VISUAL PROCESSING: FEATURE DETECTION OR SPATIAL FILTERING?

Russell L. DeValois

University of California, Berkeley

The visual system has been extensively studied for more than a century with both psychophysical and physiological techniques. This research has provided us with detailed knowledge concerning many aspects of how the early stages of the system operate. We understand (in considerable detail) the optics of the eye, the nature of the transformation of light into neural activity, and the first stages of the processing of this neural information. Not only do we now have much information on retinal processing, but also on the nature of the transformations at the first two central levels (the lateral geniculate nucleus and the striate cortex), although what happens to visual information beyond that in later cortical areas is still only dimly understood.

Despite having considerable information about the details of the early visual processing, visual scientists by no means agree upon the grand outline of what the system is accomplishing at these levels. It is clear that the ultimate objective of the visual process is to allow us to perceive and identify visual objects. There are many problems the system must solve on the way to this goal -- such as, for instance, gain control mechanisms to deal with widely varying light levels -- and certain intermediate analyses which must be done, but all of this must be leading to a mechanism for dealing with the spatial variations in the environment which we recognize as visual objects.

For some years the dominant theory of how spatial information was processed early in the visual system was that individual cells acted as feature detectors. The first (and most extreme) statement of this position was in the very influential and provocative paper by Lettvin, Maturana, Pitts, and McCulloch (1959). They categorized frog retinal ganglion cells into a few distinct varieties, each of which was detecting some fairly complex aspect of the natural environment. The most famous of their cell types were the bug-detectors, which fired only to the presence of a moving small black object of about the size of a bug. No one has ever suggested this degree of specificity (or the accompanying complexity of circuitry doubtless required) at early levels of the mammalian visual system, but related models have been developed based upon detectors for bars and edges. These models took their impetus from the pioneering studies of Hubel and Wiesel (1959, 1962, 1968), who reported the presence, in cat and monkey striate cortex, of cells which responded optimally to bars of a particular width, or to edges. These models resemble the feature-detector model of Lettvin et al. in postulating units which respond selectively to certain semi-naturalistic features of the environment (and presumably quite complex non-linear processing to produce such specificities).

The bar-and-edge theories of visual processing have had the additional characteristic of a hierarchical structure, based again upon Hubel and Wiesel's theories. They proposed that the various cell types they found in the cortex were organized in a hierarchical manner, with nonoriented cells feeding into simple cells, these in turn to complex cells, and complex to hypercomplex cells. The cells at successive levels would in general become increasingly specific for bar width and orientation. The resultant output from the striate would then be cells quite specific for bars of a particular width, length and orientation or for oriented edges. The expectation was that the outputs of these cells would at subsequent levels be combined to produce cells selective for combinations of bars and/or edges. Hubel and Wiesel (1965) in fact reported cells selective for angles, in areas 18 and 19.

This picture of striate organization, and the models of visual processing of spatial information based upon it, have been questioned recently on two grounds. One is by the increasing physiological evidence against a strictly hierarchical arrangement of cells in striate cortex; the other is an entirely different model of early visual processsing put forth by Campbell and Robson (1968) together with the now considerable psychophysical and physiological evidence for it.

Campbell and Robson (1968) proposed that the visual system may be operating in a fairly linear manner (rather than the very non-linear processing implied by a naturalistic feature-detector), with multiple spatial frequency channels. Suppose, for the moment, that cortical simple cells have the receptive field (RF) shape classically attributed to them: an excitatory center and one or two powerful antagonistic flanks (many cells in fact have further, multiple side-bands). Linear summation of excitation and inhibition within such a RF would produce a cell selective to a limited range of spatial frequencies. If the RFs of cortical cells picking up from a given retinal area varied considerably in size, there would be cells responsive to each of many different spatial-frequency ranges; that is, there would be multiple spatial-frequency channels. It can be seen that this model of early visual processing of spatial information is similar to the conception we have had since the time of Helmholtz as to how the auditory system deals with temporal information. Helmholtz suggested that the complex incoming auditory wave was Fourier analyzed in the cochlea into different sine-wave frequency components; this was presumably accomplished by virtue of the fact that the different portions of the cochlea responded preferentially to different temporal frequencies. He thus in effect proposed multiple temporal-frequency channels in the auditory system, and linear filtering of the auditory information.

One of the complexities of vision is that whereas the temporal information to the auditory system is one-dimensional, the retinal image is two-dimensional. Although Campbell and Robson (1968) only considered the processing of one-dimensionally-varying stimuli, their approach can readily be expanded to a consideration of two-dimensional patterns. The additional factor required to deal with two-dimensionally-varying patterns is in fact just that distinctive characteristic which Hubel and Wiesel found cortical cells to have, namely orientational selectivity. Spatially filtering two dimensional patterns would require multiple spatial channels at each of a number of orientations.

The physiological evidence we have been obtaining in our recent studies of the properties of monkey and cat cortical cells strongly supports the spatial-filtering model of early visual processing, as against the Hubel-Wiesel bar-and-edge detector model. I would like to summarize some of the evidence against a hierarchical bar-and-edge model, and then present some of our data supporting the notion that the cortical cells act as two-dimensional spatial-frequency filters.

Problems with the classical hierarchical model

Several lines of physiological evidence have raised doubt about a strict hierarchical organization of cells in the striate cortex, as postulated by Hubel and Wiesel, Stone (1972) and Hoffmann and Stone (1971) showed that complex as well as simple cells receive a monosynaptic input from the geniculate. Furthermore, the latency to electrical stimulation is even somewhat shorter in complex than in simple cells. Both of these findings are incompatible with the Hubel-Wiesel model of simple cells feeding into complex. Movshon (1975) also found that some complex cells have temporal properties never seen among simple cells, which could not occur of course if complex cells received an input only from simple cells. Simple cells on the average are also somewhat more narrowly tuned to both orientation and spatial frequency than are complex cells, a fact which is not logically incompatible with the Hubel-Wiesel model but would be rather unlikely in that framework.

Perhaps the most powerful argument against a strict hierarchical arrangement of simple to complex cells, with the output of the latter constituting the sole output from the cortex, is that simple cells have certain characteristics not shared by complex cells, but which are certainly represented in visual perception. Specifically, simple but not complex cells are color- and phase-specific. A complex cell can detect a luminance-varying bar, but responds equivalently to a white or a black bar in the same location. A color-sensitive complex cell can detect a color-varying bar, but again responds equivalently to a red bar or a green bar in the same location. In either situation, a simple cell would give quite different responses: if it fired to a white bar it would inhibit to a black, and correspondingly to red and green bars. Since we can obviously perceptually distinguish white from black and red from green, there must be outputs to later levels from simple as well as from complex cells.

Doubts about a model are always strengthened if reasonable alternatives can be supported. Such an alternative is suggested in the resemblance between simple and complex cortical cells and retinal X and Y cells, as defined by Enroth-Cugell and Robson (1966). The same tests which they found divided retinal ganglion cells into two groups, the X and the Y cells, divides the cortical population in two; the cortical cells which respond similarly to retinal X cells would be classified as simple by Hubel and Wiesel, and the Y cells as complex. This of course suggests a parallel organization of two varieties of cells from the retina through the striate cortex rather than an intracortical hierarchy.

It is not necessary, of course, that the cortical arrangement be either strictly hierarchical or totally in parallel. While simple and compelx cells greatly resemble retinal X and Y cells, respectively, the proportion of complex cells in the cortex (roughly 50%, by most estimates) is far higher than that of Y cells in the retina (perhaps 3%), suggesting that at least some complex cells are 'built' within the cortex from simple cells. Furthermore, we have pointed out (DeValois et al, 1979) that the responses of complex cells to checkerboard patterns would be hard to explain if they received only a Y-cell input, but would be expected if they were built up from simple cells. The critical fact remains, however, that <u>both</u> simple and complex cells must project on to later centers, since the characteristics of both (and in particular of simple cells) are reflected in our visual percepts.

While these various experiments raise some considerable doubts about the presence of a strict hierarchical organization within the striate cortex, they do not bear direclty on the main question raised in the title: is the early visual organization through the striate cortex best described as analyzing the visual world in terms of semi-naturalistic features such as bars and edges, or is it spatial-frequency filtering the information? In an attempt to answer that question we have carried out several physiological and psychophysical experiments which I would like to briefly describe here.

<u>Selectivity and Responsivity to Bars and Gratings</u>

If the visual system is analyzing the visual environment into bars of varying widths and orientations, it must perforce be selective for bar-width. If, on the other hand, the cells are filtering the image into different spatial frequency bands, one would expect the cells to be selective for spatial frequency but <u>not</u> for bar width. A bar is a broad-band spatial stimulus containing a wide range of spatial frequencies so from a linear-filtering point of view one should expect that any cortical cell would respond to bars of almost any width. One test of these alternate models, then, would be the relative <u>selectivity</u> of cortical cells to bars vs sine-wave gratings.

A second test of these alternate models would be in the relative <u>sensitivity</u> of cells to bar and grating patterns. A conventional definition of what "feature" a cell is detecting, from the point of view of feature-detectors, is that stimulus to which a cell is most responsive. Thus a bug-detector is so named because it presumably responds optimally to a moving, bug-shaped object. Hubel and Wiesel (1962) in fact stated that cortical cells were optimally sensitive to bars and edges (although, of course, they did not examine the responses of these cells to periodic stimuli).

In recordings from monkey and cat cortical cells, Albrecht, Thorell and I have examined both of these predictions (Albrecht, 1978; DeValois, Albrecht and Thorell, 1978; Albrecht, DeValois and Thorell, 1980). To examine the relative selectivity of cortical cells to bars and gratings, we drifted bars of various widths and gratings of various spatial frequencies past the receptive fields (RFs) of both simple and complex cells. The bars and gratings were equated for contrast, and patterns at each of several contrast levels were tested. The point of interest is the extent to which the cells responded differentially to the different widths of bars, in comparison to their differential responses to gratings of different spatial frequencies.

The results were completely unequivocal: without exception the cells tested were more selective for gratings of different spatial frequencies than for bars of different widths. In Fig. 1 can be seen a comparison of the tuning of two typical cells (one from cat and one from monkey) to gratings and to bars. It can be seen that both cells are much more narrowly tuned for spatial frequency than for bar width. The spatial selectivity of a cell can be conveniently quantified by measuring the bandwidth at half amplitude of the cell's responses. In the case of gratings, for instance, a cell responds optimally to a grating of some spatial frequency, with the response falling off to gratings of either higher or lower spatial frequency. The bandwidth of the cell is the distance between the low frequency grating which produces half-maximum response and the high frequency at which the response again falls to half. Specified in octaves (a 2:1 ratio), the bandwidth of cortical cells varies from about 0.6 octaves up to about 2.5 octaves for sine-wave grating patterns with a median of about 1.4 octaves. For bars of various widths, no cell had a bandwidth of less than 1.5 octaves, and most cells in fact did not show a drop in sensitivity half even for the broadest bars tested. Figure 2 shows the bandwidths for bars and for gratings for all cells.

As pointed out above, these results are not consistent with the view that cortical cells are bar detectors. Not only should a bar-

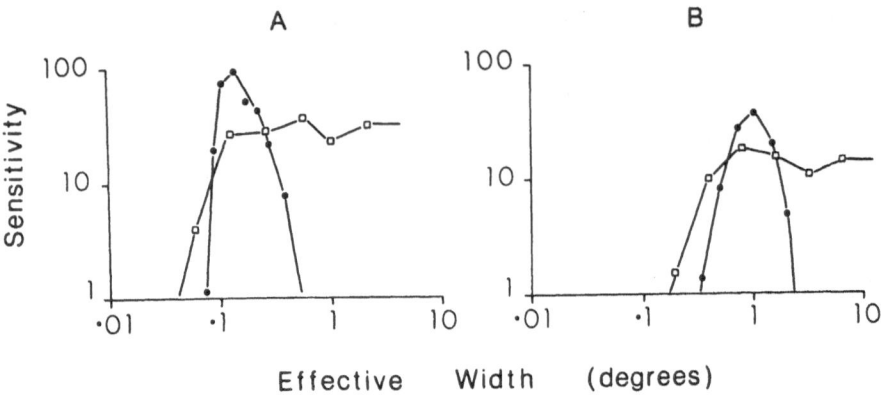

Figure 1: A comparison of spatial tuning for bars (squares) and gratings (dots) of two striate cortex cells. The cell in (A) is a monkey simple cell; in (B) a cat complex cell. In each case the contrast required to produce a change in firing rate of 1 spike/sec is plotted as a function of effective width in degrees visual angle. The effective width for gratings is taken as the width of one half-cycle. It can be seen that each of the cells shown is quite selective for gratings, but responds to bars over a wide range of bar widths. (From Albrecht, DeValois and Thorell, 1980).

detector be specific for bar width -- as these cells are not --, but it should show similar specificity for bars and gratings (which from this point of view are just multiple fuzzy bars). On the other hand, the results are qualitatively what one would expect if the cells were indeed acting as spatial filters, that is responding to patterns insofar as the pattern has power within the particular range of spatial frequencies to which the cell is sensitive.

One can go beyond such qualitative statements, however, to quantitative predictions. One of the strengths of the spatial-filter model of cell activity is that it allows one to make quantitative predictions. (The bar-and-edge model, on the other hand, is formulated very imprecisely indeed. Philosophers of science would rate

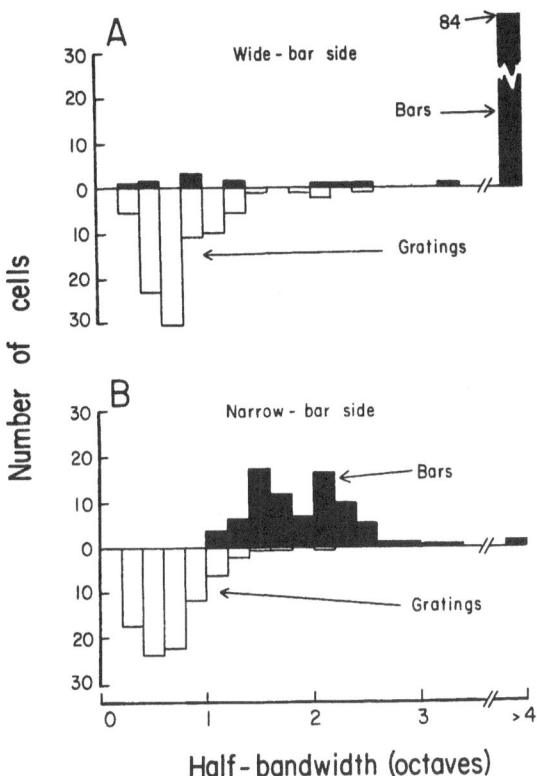

Figure 2: Bandwidths, for a sample of 96 striate neurons in cat and monkey, for bars and gratings. The half-bandwidths to either side of the peak have been measured for each cell. As can be seen, the half bandwidths for gratings are much smaller than for bars: the median half-bandwidths for gratings are 0.6 and 0.7 octaves for the two sides; for bars it is 2 octaves and more than 4 octaves for the two sides. (From Albrecht, DeValois and Thorell, 1980).

imprecision in a theory to be a major fault; in practice, an imprecisely stated theory can often be twisted around to fit new facts, however contradictory to the original formulations.) If a cell is responding as a linear spatial-frequency filter, its response to any pattern should be predictable from its sine-wave sensitivity and the spectrum of the stimulus (we are here considering just 1-dimensionally-varying stimuli). We should therefore be able to predict the responses of each cell to bars of different widths (as well as to any other one-dimensional stimulus) from the Fourier spectra of the bars. We have found that in most cases, with simple cells, one can in fact do just that with considerable accuracy (Albrecht, 1978; DeValois, Albrecht and Thorell, 1978).

In summary, not only are cortical cells much more broadly tuned for

bars of various widths than for gratings of different spatial
frequencies, but one can accurately predict from the Fourier spectra
just how large the relative responses to the various bars should be.

The data collected in these experiments (Albrecht, 1978; DeValois,
Albrecht and Thorell, 1978; Albrecht, DeValois and Thorell, 1980)
allow us to make one further comparison between the bar-detector model
and the spatial-filter model, and that is with respect to the relative
sensitivity or responsivity of the cells to bars and gratings. That
is, we can examine which of these patterns is in fact the optimal
stimulus for the cells. What we found is that when bars and spatially
delimited gratings are equated in contrast (Michelson contrast of max-
min/max+min, which is equivalent to Weber contrast $\Delta I/I$), the majority
of both simple and complex cells are not only much more selective for
gratings than for bars but are more responsive to gratings as well.
They respond better and have a higher contrast sensitivity for
gratings of the optimal spatial frequency than for bars of the optimal
bar width.

Receptive field structure of cortical cells

That sine-wave gratings are better than bars as stimuli for
cortical cells is in fact consistent with the structure of simple cell
RFs. A simple cell with a classical symmetrical RF is often described
as having an excitatory center and inhibitory flanks. This is
something of a misnomer. The RF is more accurately characterized (and
was in fact so described by Hubel and Wiesel, 1962) as having an
excitatory center and antagonistic flanks. That is, the flanks are of
opposite sign as the center, being excitatory to a stimulus of the
opposite type from that which excites the center. For example, if the
center of the RF is excited by a white bar, the flanks fire to black
bars (as well as being inhibited by white bars). The point, with
respect to the argument of whether such cells are well described as
bar-detectors, is that the optimal stimulus for such a cell is not a
single white bar. It fires quite as well to two black bars, and best
of all to a combination of these, that is, to a white bar flanked by
two black bars, or 1.5 cycles of a grating.

The assertions above are based on tests we (DeValois, Thorell and
Albrecht, in preparation) have carried out on monkey and cat cortical
cells to directly assess their responses to various numbers of cycles

of a grating. The procedure was to drift across the RF or present in counter-phase flicker various number of cycles of a grating of optimal spatial frequency, all the patterns being centered on the cell's RF. In carrying out such experiments, we have never found a cell which responded bet to only 1/2 cycle of a grating, that is, to a single bar. For many, the optimal stimulus is 1 or 1.5 cycles, as one would predict from the classical asymmetrical or symmetrical RF, respectively. About half of the cells, however, respond better to an even larger number of cycles than this, up to a maximum of 5 cycles, in our sample. Maffei and Fiorentini (1976) have also obtained similar results. We find that cells for which a larger number of cycles are optimal tend to have more narrow spatial-frequency tuning than the others, with bandwidths of one octave or less. In many of such narrowly-tuned cells, mapping the RF with a bar flashed in different loci reveals multiple side-bands which presumably account for their preference for more repetitive patterns (Albrecht, 1978). All of this is quite consistent with the view that cortical cells act as spatial-frequency filters, but quite inconsistent with their being "bar-detectors".

Cortical cells as 2-dimensional filters

In other experiments, we have obtained direct evidence that the cells are acting as 2-dimensional spatial-frequency filters. For a cell to serve as a 2-dimensional spatial filter it should have both spatial-frequency and orientation specificity. This is, of course, just what differentiates cortical from geniculate cells. Insofar as it is linear, a cell which responds to just a narrow range of spatial frequencies and orientations would have what might be termed a limited 2-dimensional Fourier response area, a restricted part of the 2-dimensional spectrum to which it would be sensitive. That is, it should (in the absence of significant interactions or other non-linearities) respond to various 2-dimensionally-varying patterns only insofar as they have power within this 2-dimensional Fourier spectral region.

To test whether cortical cells can be accurately described as 2-dimensional spatial filters, we (DeValois, DeValois and Yund, 1979) examined the responses of a sample of both simple and complex cortical cells in both cat and monkey to various simple 2-dimensionally-varying patterns. The question was whether or not we could accurately predict the responses of the cells to these various patterns from the 2-dimensional Fourier spectra of the patterns, knowing just the sensitivity of the cells to sine-wave gratings of various spatial frequencies and to gratings of various orientations.

As stimulus patterns for these tests we chose checkerboards of various length/width ratios, and a plaid. In addition to being simple 2-dimensional stimuli, these patterns have the virtue of providing critical tests of whether the Fourier spectra, or the edges of patterns are the critical determinants of the cells' responsiveness. (Bishop, Coombs and Henry, (1971) for instance, quite explicitly and emphatically state that cells respond only to the edges of patterns.) If a vertical pattern, for instance, is turned into a checkerboard its edges are still vertical (there are, of course, additional horizontal edges). As pointed out by Kelly (1976), the Fourier fundamentals of the checkerboard, however, are not at the same orientation as those of the vertical grating, but rather are displaced 45 deg. to either side. (The higher harmonics of the checkerboard have still other orientations.) A cell should thus show the same orientation tuning to gratings and checkerboards if only edge-orientation is critical, whereas the grating and checkerboard tuning should be 45 deg. apart if the orientations of the Fourier fundamentals are the critical determinants of the response.

We found (DeValois et al., 1979) that without exception the orientation tuning of both simple and complex cortical cells was predictable from the Fourier spectra of the patterns, not from the edges. In Fig. 3, for instance, are shown the orientation tuning functions for two cells (one simple, one complex) to gratings and to checkerboards. In each case, the orientation tuning was determined for the optimal spatial-frequency grating, then the grating was turned into a checkerboard and the orientation tuning tested again. If the cells are responding to the edges of the patterns, they should show the same orientation-tuning for both these stimuli. It can be seen, however, that in each case the cell gave no response at all to a checkerboard in the grating orientation. Rather, the optimal checkerboard orientation was precisely 45 deg. to either side of the best grating orientation, as predicted from the Fourier spectra of these two patterns.

In Fig. 4 are shown the results from a cell tested with a grating and with checkerboards of three different length-width ratios. It can

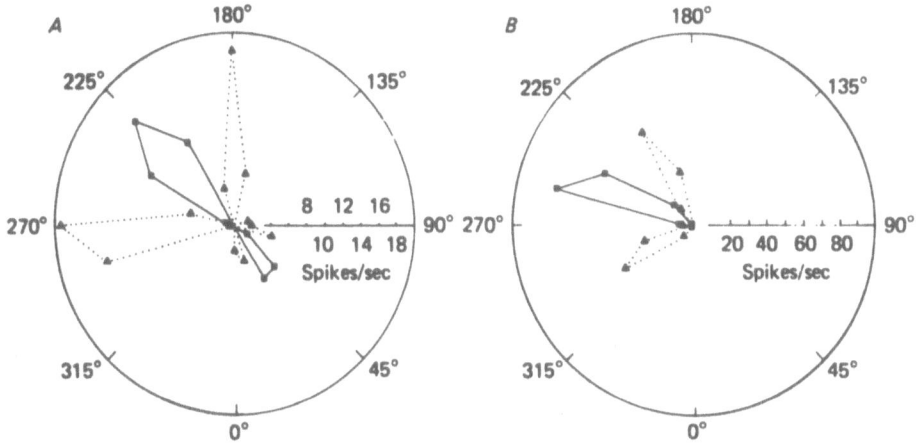

Figure 3: Orientation tuning for gratings (square and solid line) and checkerboard (triangle and dotted line) patterns of two cortical cells. The orientations plotted are those of the edges of the patterns. Note that cell A responded best to a grating of 225 deg., but gave no response to a checkerboard in this orientation; instead its tuning to the checkerboard was shifted 45 deg. to either side. The same is true for the cell shown in B. (From DeValois, DeValois and Yund, 1979).

be seen that each of these patterns produced a different orientation tuning form the cell when the orientations were specified by the edge orientation. It the "orientation of the pattern" plotted is that of the orientation of the Fourier fundamentals, however, then the cell can be seen to show exactly the same tuning in each case. So knowing the orientation of the Fourier fundamental allows one to predict precisely the optimal orientation of all of these various patterns, whereas edge orientaiton does not.

The spatial filtering and the edge-response models also make differential predictions about the optimal size of grating and checkerboard patterns. If bar-width or edge-to-edge distance is the critical determinant of optimal size for a cortical cell, then the

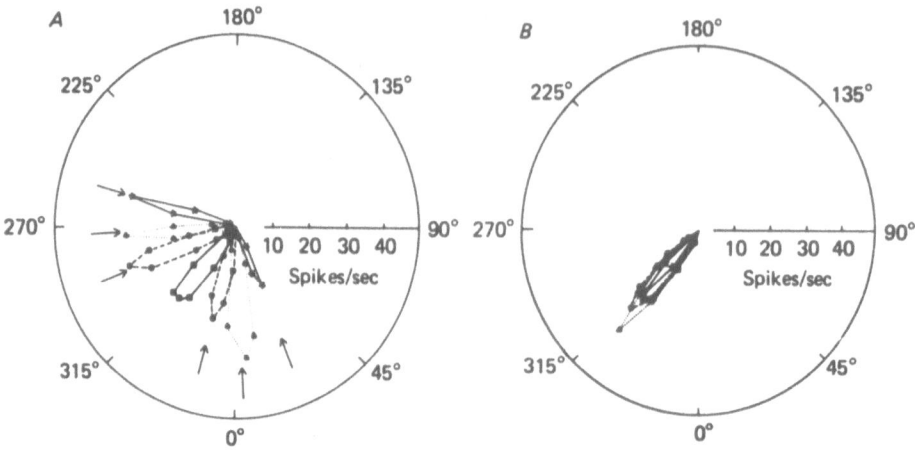

Figure 4: Orientation tunings for a cat simple cell to gratings and three different checkerboards of various length/width ratio. In A the data are plotted against the orientation of the edges of the pattern. It can be seen that the cell was tuned to a grating of 315 deg., but responded best to checkerboards of various other orientations, depending on the length/width ratio. Arrows show where the response peaks for each checkerboard would be expected to lie, from the orientation of the Fourier fundamentals. In panel B these same data are replotted, but with orientation defined as the orientation of the Fourier fundamentals, not that of the edges. It can be seen that now the data all align precisely. It is thus clear that the cell's orientation tuning is predictable not from edge-orientation, but from the orientation of the Fourier fundamentals. (From DeValois, DeValois and Yund, 1979).

cell should respond best to gratings and checkerboards of the same bar and check widths. If the spatial frequencies of the Fourier fundamentals are the critical determinants, however, a different prediction is to be made: the fundamentals of a checkerboard are shifted by 2 in spatial frequency with respect to a grating of the same bar width. A cell acting as a 2-dimensional spatial filter should thus prefer a checkerboard which has checks 1.4 times the width of the optimal-width grating.

We have measured the contrast sensitivity of cells to gratings and checkerboards of a wide range of different spatial frequencies or sizes. In each cell tested, we found that the Fourier prediction held: the contrast sensitivity function of the cell for checkerboards was shifted with respect to that for gratings by just the amount predicted from their respective Fourier spectra.

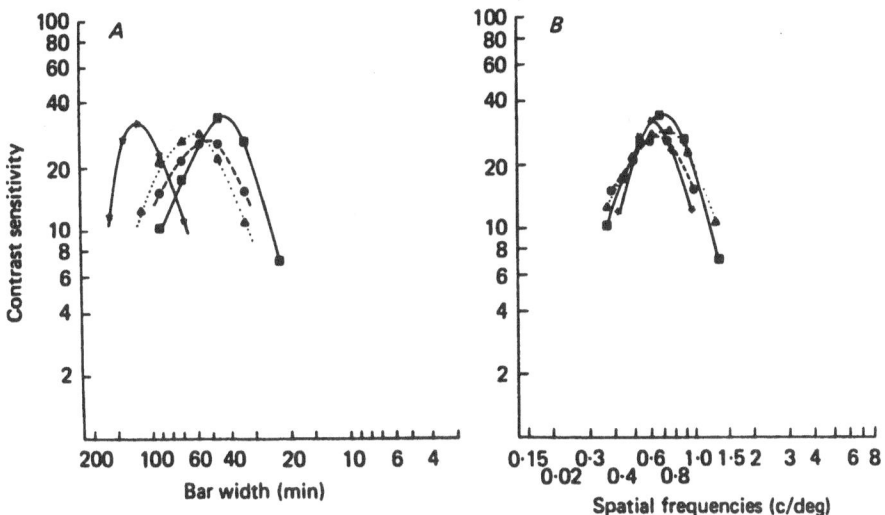

Figure 5: Spatial tuning of a cortical simple cell to gratings (square and solid line) and various checkerboards. In A the contrast sensitivity is plotted against effective bar width, and it can be seen that the peak tuning varied from pattern to pattern. When the same data are replotted against spatial frequency of the Fourier fundamental (B), the data for the various patterns coincide. Thus the spatial tuning of cells to various patterns is not predictable from bar width but is from the spatial frequency of the Fourier components. (From DeValois, DeValois and Yund, 1979).

In Fig. 5 are shown the results of such a test in which contrast sensitivities were determined for gratings, regular checkerboards and 2/1 and 0.5/1 checkerboards as well. It can be seen that, when plotted against bar and check width, the peak contrast sensitivities occur at various different widths. But when the same data are plotted against the fundamental spatial frequency of the various patterns, the contrast sensitivity functions all coincide. In other words, knowing just the check widths does not allow one to predict the optimal size pattern, but knowing the spatial frequencies of the fundamentals of the various patterns does.

One other 2-dimensional pattern that we used was a plaid, which appears quite similar to a checkerboard in appearance: the plaid has exactly the same orientations of both vertical and horizontal edges, and exactly the same bar widths as a checkerboard. Its Fourier spectrum, however, is quite different (the checkerboard is the optical product of two orthogonal gratings; the plaid, their optical sum). We found that the orientation tuning of cells to the plaid was exaclty that expected from the Fourier spectrum, not that expected from the edge-orientation and bar-width similarity to the checkerboard.

A square-wave grating, and the checkerboards made from it not only have Fourier fundamentals but higher harmonics as well. It is here, perhaps, that the edge-and-bar model and the spatial-frequency filtering conceptions diverge most drastically: the former has no place in it for higher harmonic components -- a square-wave grating is merely thought of as a series of bars of patricular width and orientation, rather than being made up of a sum of gratings of different frequencies. Examing possible responses to higher harmonics, then, provides still another means of assessing these two alternate views of the function of cortical RFs. If the cells are responding to the 2-dimensional spectrum of the stimulus, one should under appropriate conditions see responses to the higher harmonics as well as those responses to the fundamentals discussed above. In other words, if the cortical cells are acting as spatial filters, they should be decomposing complex patterns into their spatial frequency components. We (DeValois et al., 1979) have tested some of these predictions.

A square-wave grating has a fundamental spatial frequency and a third harmonic of 1/3 the amplitude (as well as the other odd harmonics). A cell which responds optimally to a grating of some spatial frequency should, therefore, also be expected to show some response to a grating of 1/3 the optimal spatial frequency (three times optimal bar width) since that pattern would have its third harmonic at the cell's preferred response range. We have found that this in fact occurs: one sees a secondary rise in a complex cell's sensitivity curve corresponding to the location of the third harmonic.

In addition, a narrowly-tuned simple cell will actually give a third-harmonic response to such a pattern, i.e., it fires three times to each transition of the pattern across the cell's RF.

Comparing the responses of cells to gratings and checkerboards, furthermore, allows one an additional test of whether there are responses to higher harmonics as predicted from a spatial-filtering model. The various harmonics of a square wave are all at the same orientation; those of a checkerboard, however, are at different orientations from those of the fundamentals. Therefore one would expect a cell to show the same orientation tuning to an optimum size grating (to whose fundamental the cell is presumably responding) and one of three-times that size (to whose third-harmonic it should now be responding). But if the same test is carried out with checkerboards of two differnt sizes, the orientation tunings should be quite different: the checkerboard fundamentals are 45 deg. from the grating fundamental, but the checkerboard third harmonics are 18 and 72 deg. away. In Fig. 6 can be seen the results from a complex cell tested with gratings and checkerboards of these two different sizes. It can be seen that the orientation tuning curves of the cell to the two sized gratings are the same, but that to the two checkerboards of different sizes differ by just the amount predicted from the Fourier spectra of the two patterns.

Prediction of random-dot adaptation effects

Considering the ensemble of cortical cells within a region as doing a 2-dimensional spatial frequency filtering of the visual information in a section of the visual world leads to various psychophysical predictions. In this regard, I would like to briefly discuss two experiments done recently in our laboratory.

One experiment (K.K. DeValois and E. Switkes, 1980) involved adaptation to random dot patterns. A random dot pattern contains energy at all spatial frequencies and orientations, that is, at all 2-dimensional spatial frequencies. However, if each dot is paired with another dot displaced laterally from it, the resultant pattern has a spatial-frequency spectrum which is periodic in one orientation, that is, there is energy in some frequency bands but not in others. The spacing of the periodicity in spatial frequency is determined by the distance between the dot-pairs. A dot-pair pattern has no perceptual resemblance to a sine-wave grating; intuitively, one would not expect any cross-adaptation effects. Furthermore, considering how dot-pairs

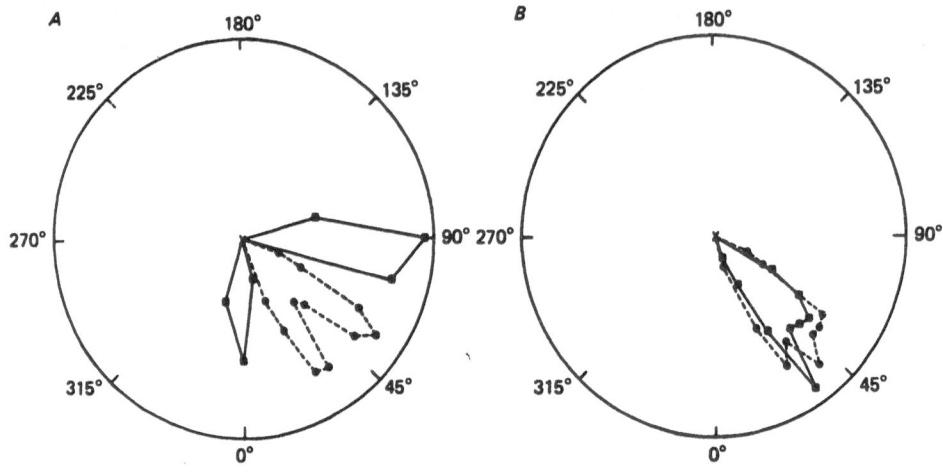

Figure 6: Responses of a cat complex cell to a square-wave grating (B) and a checkerboard (A) of different bar/check widths. In square and solid lines are the responses when the patterns were of optimal size (and the spatial frequency of the Fourier fundamentals coincided with the cell's peak tuning). In circle and dotted line are the responses to patterns three times this size, in which not the fundamentals but the third harmonics of the patterns were at the cell's peak tuning. In the case of a square-wave grating, the orientation of the Fourier fundamental and the third harmonic are the same, and it can be seen (B) that the cell showed the same orientation tuning to these patterns. The orientation of the Fourier fundamental and the third harmonic of a checkerboard, on the other hand, are different (shifted 45 deg. and 16 deg., respectively, with respect to the grating orientation), and it can be seen that the cell showed different tuning to these two patterns. So not only the tuning of cells to the fundamentals but also the higher harmonics of patterns are predictable from the Fourier spectra of the patterns. (From DeValois, DeValois and Yund, 1979).

might stimulate classic cortical cell receptive fields of differing orientations would lead one, if anything, to what turns out to be the wrong prediction. However, insofar as cortical cells act as 2-dimensional spatial frequency filters, one would expect that cells tuned to specific spatial frequencies and orientations would be adapted by a particular paired-dot pattern, and others not. Testing the sensitivity to gratings of different spatial frequencies and orientations before and after adaptation to such a dot pattern allows one to test this prediction.

The rather surprising result (K.K. DeValois and Switkes, 1980) -- surprising because it is so non-intuitive -- was that adaptation to the dot pattern did indeed produce differential effects on the detection of gratings of different spatial frequencies and orientations. In the orientation at which the power in the dot pattern is periodic, a corresponding variation was found in the adaptation effects: the rise in the adaptation effect follows closely the rise in the power spectrum and could be manipulated in a predictable way by varying the dot separation. This could only be predicted by considering the spatial frequency content of the pattern and the known characteristics of the spatial frequency channels.

At the orthogonal orientation, where the paired-random-dot pattern has uniform power at all spatial frequencies, no significant adaptational effects were found, although the power at any spatial frequency was at least as high as that for the orthogonal orientation. This result suggest inhibitory interactions among spatial frequency channels, for which other evidence (K.K. DeValois, 1977) has also been found.

In summary, then, this experiment (K.K. DeValois and Switkes, 1980) provides powerful support for a 2-dimensional spatial frequency filtering account of cortical function, although the non-linearity of inhibitory interactions must also be taken into account.

Phase vs positional sensitivity in movement detection

A second set of psychophysical experiments we have been doing recently which bear on the general question of the adequacy of a spatial-filtering account of cortical function, involved a study of the detection of movement of grating patterns of various spatial frequencies.

It is clear from our recording experiments that simple cortical cells are sensitive to changes in phase rather than changes in position. A simple cell gives its maximum response to a grating in a particular location with respect to its RF. To this same grating shifted 90 deg. to either side, the cell will give no response, and to one shifted in phase by 180 deg. the cell will be inhibited. For any given cell this relation could also be stated in terms of positional changes: it gives its maximum response to the grating in a particular position and no response to the pattern at some other position x minutes of arc away, etc. However, consider the fact that cortical cells dealing with any visual locus are tuned to a wide range of spatial frequencies -- over about 4 octaves in the foveal projection area (DeValois, Albrecht and Thorell, 1981). Each of these will show a response shift from maximum to minimum for a 90 deg. phase change, but this phase change will result from quite different positional changes for the various cells, depending on their peak spatial frequency tuning. For instance, for a cell tuned to 1 c/deg., a 90 deg. phase change is a shift of 15 minutes of arc; for a cell tuned to 10 c/deg., however, the same 90 deg. phase change is only 1.5 minutes of arc. So for the population as a whole, phase shifts and positional shifts are by no means equivalent, and the physiological data suggest that (at the striate level, at least) it is phase rather than position which would be the relevant variable.

One would expect on these grounds that we should be able to detect a certain constant phase shift of gratings of different spatial frequencies, and thus have quite different positional sensitivities. On the other hand, the real world is so arranged that position rather than phase must be the important variable. Objects made up of multiple frequencies move as a whole over constant distances, during which the different frequency components in the pattern are moving over quite different phase angles. So in dealing with the real world, one would think that the system would have to be arranged to deal with position not phase.

We (DeValois and DeValois, in preparation) examined this by testing human observer's ability to detect either lateral movement or movement in depth of gratings of different spatial frequencies. The results indicate that at low spatial frequencies, one has a constant phase sensitivity, as would be expected from the striate physiology, as discussed above, but that at high frequencies one has a constant positional sensitivity.

Most intriguingly, the results can be precisely modeled by the sum of a phase term and a positional term. That is, to detect a change requires a certain phase shift in the stimulus plus a certain absolute change in position. At low spatial frequencies the phase term is large and the positional term relatively insignificant, so one has essentially constant phase sensitivity. At high spatial frequencies, the positional term is much larger than the phase term, so one has virtually constant positional sensitivity. Such a simple sum of a phase term and a positional term suggests two successive stages of processing. One stage (presumably the striate cortex, judging from the current physiological evidence summarized above) would involve cells operating in the spatial-frequency domain with a constant phase sensitivity; a second stage (presumably operating later) would convert the processing to position, to deal with the nature of the real world. This could be readily done by weighting the contributions of cells tuned to different spatial frequencies in inverse relation to their peak spatial frequency tuning points, cells tuned to low spatial frequencies being given much more weight than those tuned to high frequencies.

Summary

In studies of the responses of simple and complex cortical cells in both monkey and cat we have found that the cells are more responsive to and much more narrowly tuned for gratings of different spatial frequency than for bars of different widths. This indicates that the cells are better described as spatial filters than as bar-detectors or edge-detectors. That they can indeed be accurately described as 2-dimensional spatial filters is indicated by the fact that we found we could accurately predict both the orientation and the spatial-frequency tuning of cells to various grating, checkerboard and plaid patterns from their Fourier spectra. Of particular significance is that we obtained direct evidence from the firing patterns of single cells, and indirect evidence from the orientation tuning of both simple and complex cells for their dissecting, square-wave gratings and the checkerboard patterns into their harmonic components. (See also: Albrecht and DeValois, 1981)

In two psychophysical experiments discussed, further evidence for spatial-frequency processing was reported. The adaptation effects of random-dot patterns on gratings are predictable from their Fourier spectra (but inhibitory interactions must also be considered), although these patterns bear no perceptual resemblance to each other in the space domain. A phase (as opposed to positional) term was found, as expected from the spatial-frequency characteristics of cells, in an observer's sensitivity to moving gratings.

References

Albrecht, D.G. (1978) Analysis of Visual Form Doctoral dissertation, University of California, Berkeley. 1-251.

Albrecht, D.G. and DeValois, R.L. (1981) Striate cortex responses to periodic patterns with and without the fundamental harmonics. Journal of Physiology (in press).

Albrecht, D.G., DeValois, R.L., Thorell, L.G. (1980) Visual cortical neurons: Are bars or gratings the optimal stimuli? Science 207: 88-90.

Bishop, P.O., Coombs, J.S., Henry, G.H. (1971) Responses to visual contours: spatio-temporal aspects of excitation in the receptive fields of simple striate neurones. Journal of Physiology 219: 625-657.

Campbell, F.W., Robson, J.G. (1968) Application of Fourier analysis to the visibility of gratings. Journal of Physiology 197: 551-556.

DeValois, K.K. (1977) Spatial frequency adaptation can enhance contrast sensitivity. Vision Research 17: 1057-1065.

DeValois, K.K., DeValois, R.L., Yund, E.W. (1979) Responses of striate cortex cells to gratings and checkerboard patterns. Journal of Physiology 291: 483-505.

DeValois, K.K., and Switkes, E. (1980) Spatial frequency specific interaction of dot patterns and gratings. Proc. Nat. Acad. Sci. 77: 662-665.

DeValois, R.L., Albrecht, D.G., Thorell, L.G. (1978) Cortical cells: Bar detectors or spatial frequency filters? In Frontiers in Visual Science. Cool, S.J., Smith, E.L., Eds. Berlin:Springer. pp.544-556.

DeValois, R.L., Albrecht, D.G. and Thorell, L.G. (1981) Spatial frequency selectivity of cells in macaque visual cortex. Vision Research in press.

DeValois, R.L., Thorell, L.G., Albrecht, D.G. Responses of cortical cells to various numbers of cycles of a grating. In preparation.

Enroth-Cugell, C., Robson, J.G. (1966) The contrast sensitivity of retinal ganglion cells of the cat. Journal of Physiology 187: 517-552.

Hoffmann, K.P., Stone, J. (1971) Conduction velocity of afferents to cat visual cortex: a correlation with cortical receptive field properties. Brain Research 32: 460-466.

Hubel, D.H., Wiesel, T.N. (1959) Receptive fields of single neurones in the cat's striate cortex. Journal of Physiology 148: 574-591.

Hubel, D.H., Wiesel, T.N. (1962) Receptive fields, binocular interaction and functional architecture in the cat's visual cortex. Journal of Physiology 160: 106-154.

Hubel, D.H., Wiesel, T.N. (1965) Receptive fields and functional architecture in two nonstriate visual areas (18 and 19) of the cat. Journal of Neurophysiology 28: 229-289.

Hubel, D.H., Wiesel, T.N. (1968) Receptive fields and functional architecture of monkey striate cortex. Journal of Physiology 195: 215-243.

Kelly, D.H. (1976) Pattern detection and the two-dimensional Fourier transform: Flickering checkerboards and chromatic mechanisms. Vision Research 16: 277-287.

Lettvin, J.Y., Maturana, H.R., McCulloch, W.S., Pitts, W.H. (1959) What the frog's eye tells the frog's brain. Proc. Insti. Radio Eng. 47: 1940-1951.

Maffei, L., Fiorentini, A. (1976) The unresponsive regions of visual cortical receptive fields. Vision Research 16: 1131-1139.

Movshon, J.A. (1975) The velocity tuning of single units in cat striate cortex. Journal of Physiology 249: 445-468.

Schiller, P.H., Finlay, B.L., Volman, S.F. (1976) Quantitative studies of single-cell properties in monkey striate cortex: III. Spatial frequency. J. Neurophysiol. 39: 1334-1351.

Stone, J. (1972) Morphology and physiology of the geniculo-cortical synapse in the cat: quesiton of parallel input to the striate cortex. Invest. Ophthal. 11: 338-346.

ON A FILTER APPROACH TO UNDERSTANDING THE PERCEPTION OF VISUAL FORM

by

Arthur P. Ginsburg
Aviation Vision Laboratory
Air Force Aerospace Medical Research Laboratory
Wright-Patterson Air Force Base, Ohio 45433

Abstract

One fundamental problem facing vision research is the lack of a physics of form that can describe (in the same mathematical language) both object information and the visual mechanisms. This problem has allowed a large number of competing theories of vision to exist, each offering a different solution to common visual problems. However, criteria can be established to help evaluate the various theories of vision. A physics of form is described, within the context of filtering, that is shown to satisfy reasonable criteria. Demonstrations show that the filtering approach is able to quantify form, solve certain aspects of form generalization and help explain how a wide range of information contained in complex objects can be extracted by a small numbered bank of spatial channels.

The central problem of form perception is to understand how the observer obtains information from objects. What are the properties of spatial distributions of objects that provide meaning? How do visual mechanisms extract those meaningful properties? Although these questions have been grappled with for many years by vision scientists, and in spite of our rapidly increasing knowledge about visual science, no general theory of form perception exists. Many theories of vision have been proposed, but none have stood the test of time. No visual theory seems to have captured the essence of scientific understanding: parsimonious, quantitative, predictive power. Longstanding problems suggest that the basic assumptions must be reexamined. One basic assumption in vision research is that (a) object information, (b) psychophysical and neurological data and mechanisms, and (c) visual response can all be described independent of one another. This assumption is questioned: it is argued that a common mathematical framework is needed to help unify and provide perspective for the plethora of vision data that is currently being generated. A framework using the concept of filtering is suggested to help solve that problem.

There are three general parts to understanding form perception: object, visual processing and response. Unfortunately, little research has attempted to create a framework for specifying form information conveyed by an object or pattern. Here an object is defined as connected spatial features. A pattern is a disconnected set of spatial features. The shape information extracted from objects and patterns is form. An important aspect of form perception is the abstraction of similar information from sometimes dissimilar pattern features. Letters having similar form can be classified as the same even though they are created from discriminably different fonts. Thus an understanding of form perception requires an understanding of the visual mechanisms responsible for the abstraction process: the selection of certain features of the object or pattern. In this context, abstraction is synonymous to filtering: the selection of a certain range along a particular physical dimension of the object or pattern. If this premise is accepted, then one way to possibly unify the problems, concepts, data and theories of form perception is within the context of filtering. It is suggested that the lack of a physics for specifying form is a major reason for our lack of understanding form perception. Theories of vision should be able to specify form in the same language used for specifying visual processing.

The fundamental purpose of the visual system is to interpret or process information in the external world relevant to the needs of the organism. The mechanisms that perform that objective include optics, receptive fields, and cognitive strategies. The resulting visual system is so complex that a complete analysis of the overall system which includes all of its components is not possible. Therefore it becomes necessary to subdivide the complete system into smaller parts, each of which can be mathematically modeled. The necessity for subdivision of the visual system is a major problem for visual science as evidenced by the different disciplines found probing the mysteries of vision: psychophysics, neurophysiology and psychology. Each discipline, with its different emphasis, rarely deals with more than a small portion of the total visual system. That division of labor has resulted in a large number of non-coherent theories of vision that may indeed be valid at one level, but are easily faulted at another level and rarely have similar quantitative frameworks. However, the failure of any one model to predict performance at all levels and for all tasks does not mean that the model is bad or non-useful.

Although it seems in vogue in certain quarters of the visual community to dismiss a particular model because it is incomplete (for example, failing one particular "test"), that behavior indicates a lack of fundamental understanding of models, namely that a mathematical model is always an approximation. This same kind of binary thinking is also sometimes used when considering the linearity and nonlinearity of the visual system. Some researchers, upon finding some non-linearity at some level of visual processing, loudly proclaim the total non-linearity of visual processing at all levels thereby throwing out the important and useful tools of linear system analysis. Mathematical models always represent a compromise between the conflicitng requirements of accuracy and mathematical simplicity. It is the nature of the physical problem and level of understanding that dictates how this compromise is made. However, in reality it is usually the educational background and experience that typically biases that compromise by placing emphasis and analysis on one aspect of the system over another. This appears to be a major reason why different scientists create different descriptions and models for the same visual mechanisms.

Any viable theory of form perception must eventually grapple with the physics of visual mechanisms. Clearly it is the physical attributes of visual mechanisms that limit the capabilities and bound the kinds of information processing that can occur in vision. For example, no matter how well motivated or trained an individual is, visual spatial acuity is limited by optics and physiological sampling. That means the visual system can only process a certain range or size of object. (Similar logic applies to the "temporal acuity".) It is well known that the size of retinal receptive fields increases greatly from the fovea to the parafovea which limits the size of spatial information passed. What these kinds of sample limits on visual input mean (and similar arguments can be made for motion, color, brightness, depth perception, etc.), is that the visual system performs large amounts of data reduction at early stages of visual processing. One way to describe that data reduction is in terms of filtering. However, major disagreement ensues when the next step is taken, that of describing the filter properties of visual mechanisms.

The vast majority of research attempting to understand form

perception has stressed the role of "visual mechanisms" (especially spatial-analytic mechanisms); such research has resulted in the creation of a wide variety of potential mechanisms, each capable of extracting quite different properties of form. In general, two main groups of vision scientists have emerged over the last 22 years that profess different conceptual views of the roles of visual mechanisms that mediate form perception. One group views visual mechanisms as feature detectors: receptive fields that act as templates for a group of highly specific pattern properties such as bar-widths and edges. The other group, taken to its extreme, views visual mechanisms as Fourier analytic: receptive fields highly tuned to sinusoidal distributions of intensity (that is, spatial frequency).

Although the concept of feature detection has been a reasonable way of characterizing receptive field behavior in a qualitative manner, feature detection approaches have not generally encouraged quantitative descriptions of visual form. The assumption that there is a one-to-one mapping between somewhat arbitrarily chosen geometric properties of the stimulus object and the receptive field response led to a flood of feature specific classes of receptive fields and theories of vision (e.g. Barlow, Narashimhan and Rosenfeld, 1972). Unfortunately the mind-set created by directly relating geometric stimulus features to feature detection has caused many individuals to view receptive fields as bar detectors or spatial frequency grating detectors because the receptive field responds maximally to bars or sine-wave gratings. It is easy in retrospect to criticize that approach to understanding vision, but the seductive need to describe receptive field properties in terms of geometric features is quite compelling. The problem, of course, is that even a "simple" bar or dot is not a simple stimulus; it has more properties than just size (for example, edges) that could also have been specified as a feature. The reality of the situation is simply that the receptive field, like any other part of a complex information processing system, is acting as a filter by passing some range of information and rejecting others. The problem then becomes one of quantifying that range of information in terms that can, in turn, be used to quantify the size, shape, area, edges, contrast and other features that comprise the object.

Arguments for and against feature detectors and Fourier analyzers can be found in much of vision literature and do focus on some important issues about certain functions and processes of visual mechanisms. However, those same arguments can mask the commonality of those two descriptions. It should be clear that both descriptions are usually technically correct in the context used by the scientist. What will be suggested here is that both descriptions can be subsumed within the quantitative framework of spatial filtering. The power of that notion comes from forcing a shift from a qualitative description of the particular geometric properties of the object to the quantification of specific ranges of certain physical properties of the object. In other words, the concept of filtering forces the analysis of form perception into a different frame of reference where spatial information must be quantified.

When evaluating a theory of form perception, one might minimally expect the theory to be: (a) quantitative, (b) parsimonious, (c) biologically consistent and (d) testable in a manner that relates to how we use information from objects to see and act. What is needed is a coherent framework for analysis, a quantitative through-put from stimulus to response, object to performance. Unfortunately, most frameworks for understanding form perception do not satisfy the preceding criteria and (at least in my opinion) can be dismissed as serious candidates for a meaningful understanding of form perception. It is further argued that the emphasis on understanding visual mechanisms to the exclusion of understanding the need for specifying relevant information about objects and how that information is being used for form perception is also dangerous.

There is, in fact, a rich mathematical universe with which to describe visual mechanisms. The problem is the following: how is one to select relevant descriptions from such a large number of possibilities? One way out of the prevailing plethora of descriptions of visual mechanisms is to use a language for describing meaningful information about objects that produces a response from the mechanisms and then evaluate the response of the mechanisms and describe their function in terms of that information. In other words, let a physics of form guide the quantification of visual mechanisms. Let us see where this emphasis can lead to.

It is possible to develop a physics of form (a mathematical framework that can describe any object); there are several basic considerations. First it is necessary to represent an object in terms of explicit space functions whose numerical values are exactly defined. This means that quantitive descriptions of object features and/or visual mechanisms must be mathematically defined. Stating that a receptive field is a bar-detector merely because a maximum response occurs when the stimulus object is a bar is not sufficient to lead to any quantitative predictions.

The next consideration is the criteria that should govern the selection of the space functions. Although there are many ways in which this mathematical representation could be realized with the same objective, the choice of any particular method is usually dictated by mathematical considerations. Three desirable properties of space functions are linearity, finality of coefficients and minimum integral squared error of representation (see Ginsburg, 1978 for further discussion). A linear set of space functions has the important advantage of testing for superposition, whereby the response of the system to complex objects can be calculated simply by taking the sum of the responses of the system to individual simple patterns. Finality of coefficients is a property that allows the value of the coefficients (that may represent for example, form) of each space function to be determined without the need to know the values of other coefficients. In other words, each space function carries with it a discrete description of some property of the object; it is independent of the value of any other coefficient. This property can allow more space functions to be added to allow greater accuracy in the representation of the object without making any changes in the earlier coefficients. Finally, any real object stored, for example, in a digital computer (or the visual system) can be represented only by a finite set of space functions. In this case, it is important to pick a set of space functions that minimize the error of representing an object when only some and not all the terms are used. One means of evaluating the error of representations is the integral squared error. It can be shown that certain space functions not only have the property of finality of coefficients but also minimize the integral squared error of representation.

Let us review these properties of this descriptive framework for objects. First, we have a finite set of functions used to mathematically describe any object. Second we can test the response of the visual system with any one or more space functions and determine how the response to those space functions will predict the response to more complex objects. Third, these space functions will represent the actual pattern with minimum error. Of course other criteria could be used to select space functions, but that discussion would take us beyond the scope of this paper.

At this point, the already initiated reader will realize that the preceding discussion is little more than a suggestion of linear systems framework for quantifying spatial information in objects. In other words, there already exists a framework for a physics of form. Hopefully, the uninitiated reader will soon realize that there has already been considerable work and success in considering how information in spatial objects can be mathematically represented. If the reader at this time suspects a certain bias towards this approach, it is because linear systems analysis is a widely accepted and general framework used in many disciplines that has helped to understand many complex information processing systems. It would seem reasonable to suggest that this existing approach offers as much promise in helping to understand the visual system as it has in helping to understand other complex physical systems. Further, it would seem that until another approach is demonstrated to have at least this degree of analytical power (or can solve important problems that this approach cannot solve) then it can hardly claim to be useful within the constraints of the preceding goals. The bottom line is that much scientific energy expended in attempting to create new analytical frameworks without considering the already existing frameworks may be wasteful and can direct attention away from the more interesting and serious problems of form perception.

The preceding discussion about criteria used to select space functions with which to quantify information in objects does little to restrict the particular set of space functions used. The selection of space functions depends primarily upon how the object representation is to be used. In the analysis of complex physical systems, sinusoids exhibit many useful mathematical properties; they remain sinusoids upon addition, subtraction, differentiation and integration. These properties, combined with superposition, make the representation of objects as a sum of sinusoids quite convenient as well as powerful. Over the last 20 years large advances in quantifying certain aspects of vision have been made by scientists employing sinusoids and linear systems analysis primarily in terms of Fourier techniques. This approach has led to the concept of visual channels (Pantle and Sekuler, 1968; Blakemore and Campbell, 1969), contrast sensitivity functions (Campbell and Robson, 1968) quantifying certain information about form and a general theory of form perception (Ginsburg 1971, 1978, 1980; Ginsburg, Cannon, and Nelson, 1980).

If one accepts the preceding approach, then the next step, a way to quantify how visual mechanisms extract that kind of information, is straightforward. The natural relationship between the stimulus object and response can be specified in terms of filtering, the amount of spatial information that is passed by the visual mechanism in terms of the space functions, sinusoids. The relevant questions that tie the physics of form to the filtering properties of visual mechanisms include: what is the bandwidth (the range of spatial information, e.g., sizes or spatial frequencies passed by the filter), center frequency (a reference point, usually the frequency of maximum response of the filter) and the weighting function (the shape of the filter, the way it attentuates the information over the bandwidth). Further questions that can be used to relate filtering to form perception are: how do visual filters extract the forms we perceive and how do those filters make the pictures that we see? And once these questions are answered, how do cognitive processes and other visual mechanisms select and interpret information from filtered images?

Note that the approach does not attempt to determine the computation of the filtering mechanisms, but rather argues that it may be more important to know how and to what degree visual mechanisms transmit, for example, size, contrast and orientation information. It does not seem sufficient to simply assert that those kinds of physical characteristics are computed using some arbitrary values. For example, stating that receptive fields in the visual system compute certain stimulus attributes may be as unimportant as suggesting that electrons in a resistor compute relationships between voltage, current and resistance. The important relationship between voltage, current and resistance is described by Ohm's law. The filter approach attempts to search for the Ohm's law (or laws) of vision, meaningful relationships between physical properties of objects and visual mechanisms within a similar mathematically well-defined framework. The filter approach uses such mathematical tools as linear systems analysis to determine the implications of the overall and individual filtering characteristics of visual mechanisms on perception. Since much information has been gained about such biological data in the last 15 years, an understanding of how complex form information is being extracted by visual mechanisms may help create a firm foundation for understanding visual perception. Characteristics of certain visual mechanisms based on biological data and relevant to filtering are summarized by several scientists (Sekuler, 1974; Breitmeyer and Ganz, 1976; Albrecht, 1978; Braddick, Campbell and Atkinson, 1978; Albrecht, DeValois and Thorell, 1980; DeValois and DeValois, 1980; DeValois,

Albrecht and Thorell, 1981).

The filtering approach has been used by this author over the last 10 years to help understand the perception of generalized form, Gestalt principles of pattern grouping, geometric illusions, certain aspects of texture patterns and multistable figures (Ginsburg 1971, 1978). Those results suggest that the low spatial frequencies of objects provide the generalized form information and the high spatial frequencies provide the details and redundant information about form. The integration of that earlier work into current knowledge about the filtering characteristics of visual mechanisms has resulted in a different way of understanding visual processes. A few results from that research will show the power of the filter approach to help solve some longstanding problems of form perception.

One major problem mentioned earlier is the need to quantify relevant form information. A method that can be used to quantify sufficient information for the recognition of any form in terms of Fourier synthesis is shown in Figure 1. Here the letters, E and L, having equal size and stroke width, are created by successively adding 0.5 cycles per letter width (cpl). Note that the E requires 2.5 cpl as the minimum bandwidth or number of cycles for recognition, however the L requires only 1.5 cpl. It should be noted that these different bandwidths for similar size letters are not intuitively obvious from analyzing the letter features such as line length, width, or angle. Indeed there had been attempts to relate the visibility of these kinds of letters commonly found on eye charts to overall visual filter characteristics via contrast sensitivity. Those attempts failed because of the belief that the critical size dimension for the recognition of those letters was the stroke width. However, as this simple demonstration shows, letters having similar stroke width can have almost a factor of two difference in bandwidth requirements for recognition (which accounts for optometrists noting for years that a letter L is more visible than a letter E on eye charts). The added information provided by Fourier synthesis is the quantification of the spatial arrangement of those features. Using this minimum bandwidth approach to relate cycles per letter width to cycles per degree of visual angle, this author was able to offer strong predictions of visual acuity using typical eye chart letters based on individual contrast sensitivity functions and thus help establish a relationship between complex form perception and contrast sensitivity (Ginsburg 1978, 1980).

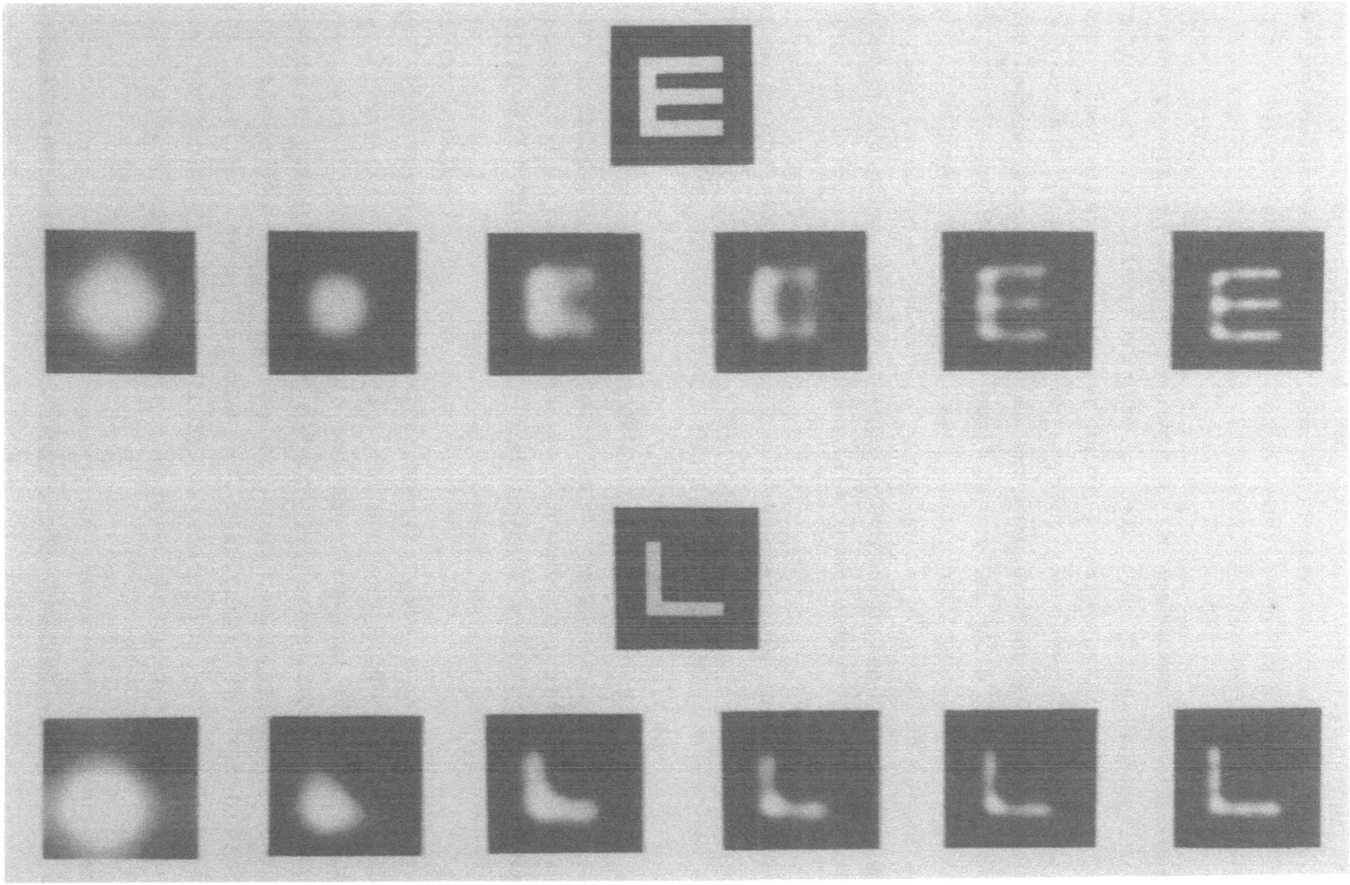

Figure 1: Fourier synthesis of letters E and L in 0.5 cycles per letter width (cpl) increments. From left to right is the original letter, 0.5, 1.0, 1.5, 2.0, 2.5, 3.0 cpl (from Ginsburg, 1978, 1980).

Another major problem hinted earlier was the capability of the visual system to abstract form from discrete patterns. Here too, some success has been achieved using the filtering approach. For example, one of the classic problems of perception the early Gestalt psychologists wrestled with was how the visual system groups discrete pattern elements into meaningful units of information. The filter approach asks what information about objects can be found in images from the physical filters known to exist in the visual system. In other words, could fundamental properties of grouping be an inadvertent concomitant of basic biological filtering? The answer appears to be yes. Earlier answers to that question showed that the same small bandwidth, one to two octaves, that captures the generalized form of many solid objects also produces solid images of generally uniform intensity distribution even from patterns that are comprised of discrete features.

Figure 2 shows examples of certain Gestalt laws of perceptual grouping that are understandable in terms of filtering: dot letter G (a), dot letter R in random noise (b), and the patterns that represent examples of grouping properties of vision - a single whole group (c), three groups due to proximity (d), two similar groups separated by a different central group (e), three groups where proximity subsumes the similarity of the pattern elements (f). The original patterns in Column 1, filtered by overall filter characteristics based on contrast sensitivity, results in the filtered images in Column 2. Column 3 contains filtered images where the fundamental frequency or first harmonic is removed to reveal forms comprised of only three low spatial frequencies. In general, the overall filtering characteristics of the visual system, under the conditions used here, simply smooth the dots. The generalized whole units of form perceived from the original pattern can be seen to exist in a range of only three spatial frequencies. Here we see simple filtering based on biological data, rather than extensive computation of pattern features, able to extract complex visual form. And, most important, the general form was captured with a minimum bandwidth, an important consideration for memory storage.

One final demonstration shows what kind of information about a complex object, a portrait, can be extracted from a bank of filters based on biological data. A portrait (filtered into eight different images using filters whose center frequencies were one octave apart and whose bandwidths were two octaves) is shown in Fig. 3. Spatial

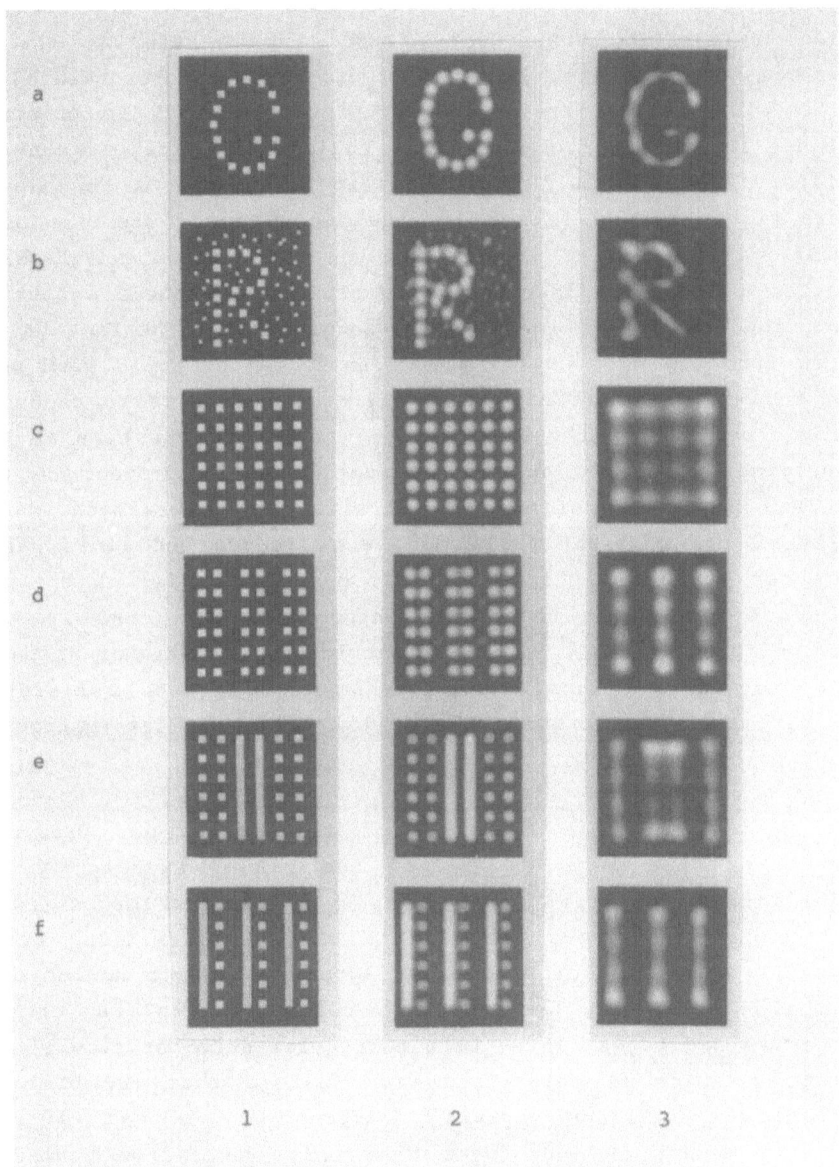

Figure 2: Examples of certain Gestalt laws of perceptual grouping: closure (a), closure and figure/ground (b), single whole group (c), three groups due to proximity (d), two similar groups separated from a different central group (e), three groups where proximity subsumes the similarity of the pattern features (f) (from Ginsburg 1978).

information transmitted by a two octave filter having a center
frequency of a 0.5 cycles per face width is shown in the picture
labeled fc=1. (The subsequent picture labels refer to the center
frequencies of the filter in terms of cycles per picture width. Cycles
per face width is half the cycles per picture width.) The existence of
an object is evident from the large regions of different contrast in
the first image (fc=1). The gross elliptical shape of an object that
appears right side up is seen in the second image (fc=2). The third
image (fc=4) definitely provides enough information to classify the
object as a face. The identification of the face needs a little more
information. The fourth image (fc=8) suggests that the face is that of
a woman from the hair style, etc. Identification would seem possible
given a limited set of similarly filtered portraits to choose from.
Clearly, the face in the fifth image (fc=16) is the same as that of
the original portrait, and identification would present no problem
given any large set of other portraits. If information about the
details of the portrait is needed, e.g., edges and lines, the hair
across her forehead, the texture of her hair, size of the pupils,
outline of her lips, then that information is found in the two
remaining images (fc=32, 64). The last image, fc=128, has virtually no
usable information. Thus, these seven images create a hierarchy of
filtered images that provide the full spectrum of information needed
for any perceptual task.

The results of this hierarchy of filtered images provide many
interesting suggestions as to how the visual system may process
spatial information. First note that seven filtered images (excluding
the last one that shows very little useful information) provide a
lexicon of spatial information that agrees well with how objects are
described: existence, general form, classification, who or what it
is, texture and edges. Thus, this particular decomposition of spatial
information which is determined from filters that are biologically
realizable, partitions spatial information in an intuitively
satisfying manner. Second, note that only one filter having a two
octave bandwidth centered at no more than 16 cpw is needed to identify
an individual face.

Note the relative contrast between different filtered portraits,
which approximates the energy of the features of the original portrait
in each bandwidth. Predominant energy can be seen to exist from
filters having 2, 4, 8, and 16 cycles per face width: the relatively
low spatial frequencies of the portrait. The high spatial frequencies

captured by the filters having higher center frequencies have greatly reduced energy. This observation further supports the notion that it is the lower spatial frequencies rather than the higher spatial frequencies that convey major information about objects. Although the higher spatial frequencies contain redundant form information, the lower spatial frequencies will provide high signal-to-noise under certain conditions. For example, the high spatial frequencies will be the first to be degraded under conditions of poor illumination, movement, viewing distance, scan structure, etc., a poor choice of information to have perception based upon in the real world.

Many people have the notion that lateral inhibition from biological receptive fields serves to "sharpen" the image. These portraits have been filtered using filters having biologically derived weighting functions. It is not evident that any edges have been sharpened in any filtered image. What is seen is enhanced contrast at the edges. No edge appears any sharper than those in the original portrait. Although it is possible that high frequencies could be enhanced to create a "sharper" image when all the channel images are combined to create the image that we see.

Finally it should be stressed that the information extracted from the portrait, from gross form to the sharpest edges did not require different filter shapes having different names. Merely changing the bandwidth, the size of the filter, allows the extraction of gross form or sharp edges; special classes of feature detectors were not needed, nor were complex computations. Similar shaped filters having a range of size of about 7 octaves are very capable of passing a wide variety of relevant form information from complex objects.

The general results of filtering objects suggest that our broad, overall filtering characteristics described from contrast sensitivity data limits the range of size of objects that can be seen. The visual processes concerned with object perception are composed of narrow-band, orientation-selective filters. These filters capture relevant form information. Thus, outputs of these individual filters appear to be presented simultaneously to create the global picture that we see.

These results are not suggested to be the only possible solution to these particular visual problems. There are many other possible solutions within and without this concept of filtering. However, these particular results are consistent with the physics of form and constraints on known biological filtering presented earlier. Whether or not this particular approach continues to be useful awaits future experimentation. Although simple filtering and Fourier techniques appear to offer solutions to the preceding problems of isolated forms, it should also be stressed at this point that these results do not imply that all aspects of spatial vision should be described in terms of Fourier techniques. There are many other serious problems of vision that may require other techniques. For example, the extraction of figures from ground, the isolation of object from background in complex scenes, may require more complex processing than just simple filtering (e.g. Ginsburg 1973, 1978). And there, such concepts of edges, boundaries, area, etc. to help segregate objects may be best described in geometric terms rather than transform coefficients. However, it does seem clear that geometric features will benefit from descriptions in terms of spatial frequency, orientation, contrast and phase in the domain of filtered images. Finally, the cognitive concepts such as "sets" and "selective attention" and "experience" need to be addressed, hopefully within biological constraints.

One final criterion that should be used to evaluate various theories of vision is the ability to predict everyday visual performance. As previously discussed, the approach of quantifying information about objects and determining their relevant bandwidth in terms of Fourier techniques does allow a relationship to be made between overall filtering characteristics of vision in terms of contrast sensitivity and, for example, visual acuity of common eyecharts. This kind of knowledge helps explain why certain people having muliple sclerosis can pass an eyechart test and yet complain about the visibility of their visual world. Those kinds of deficits appear detectable using contrast sensitivity. Further, this approach has allowed an understanding of the kinds of spatial information certain patients having low vision can see (Ginsburg, 1981a). Thus, these techniques have allowed insights into certain clinical problems that have not yielded to previous ways of looking at visual processing. In addition, these techniques have shown dramatically the inadequacy of our present methods of assessing vision in terms of visual acuity and offer a reasonable alternative if not a quantum jump to improve visual testing and selection methods. For example, large individual differences among people considered normal in terms of visual acuity have been found using contrast sensitivity techniques that correlate well with the ability to detect complex targets (Ginsburg, 1981a,b; Ginsburg, Evans, Sekuler and Harp, submitted).

Indeed, in certain cases, an individual without glasses can detect and identify low contrast larger complex targets than an individual who does not require glasses (Ginsburg, 1981a,b).

These results suggest serious investigation into contrast sensitivity techniques as a means for creating new visual standards. These new visual standards, since they are being based on real-world visual performance, promise to offer more reasonable selection criteria for a wide variety of occupations, such as pilots. In sum, not only are Fourier techniques and linear systems analysis proving powerful in being able to pry open some of the secrets of vision, they are also opening up new understanding of abnormal vision and promise to replace 100 year-old vision standards with techniques that relate well to how we go about seeing and performing in the everyday world.

In summary, this brief paper has tried to show that there exists a framework that ties together a physics of form and filtering that is relevant to a quantitative approach to understanding form perception. This framework is suggested to be broad enough to encompass many of the important issues about form perception; it may provide a common ground for an understanding of those same issues. The basic argument presented here is that in order to quantify form perception, the visual stimulus, as well as the response of the visual mechanisms, must be quantified within the same mathematical framework. The concept of filtering offers a parsimony of analysis, a coherent framework from stimulus to response. It is suggested that within the framework of filtering, there may be some unified solutions to the great mysteries of vision.

References

Albrecht, D.G. (1978) Analysis of Visual Form. Doctoral Dissertation, University of California, Berkeley.

Albrecht, D.G., DeValois, R.L., Thorell, L.G. (1980) Visual cortical neurons: are bars or gratings the optimal stimuli? Science 207: 88-90.

Barlow, H.B., Narashimhan, R., Rosenfeld, A. (1972) Visual pattern analysis in machines and animals. Science 177: 567-575.

Blakemore, C., Campbell, F.W. (1969) On the existence of neurons in the human visual system selectively sensitive to the orientation and size of retinal images. Journal of Physiology 203: 237-260.

Braddick, O.J., Campbell, F.W., Atkinson, J. (1978) Channels in vision: Basic aspects. Leibowitz, H.W., Teuber, H.L., eds. Handbook of Sensory Physiology, Vol. 8: Perception, Berlin: Springer.

Breitmeyer, B.G., Ganz, L. (1976) Implications of sustained and transient channels for theories of visual pattern masking, saccadic suppression, and information processing. Psychological Review 83: 1-36.

Campbell, F.W., Robson, J.G. (1968) Application of Fourier analysis to the visibility of gratings. Journal of Physiology 197: 551-566.

DeValois, R.L., Albrecht, D.G. and Thorell, L.G. (1981) Spatial frequency selectivity of cells in macaque visual cortex. Visual Research (in press).

DeValois, R.L., DeValois, K.K. (1980) Spatial vision. Annual Review of Psychology 31: 309-41.

Ginsburg, A.P. (1971) Psychological correlates of a model of the human visual system. Proc., 1971 National Aerospace Electronics Conference (NAECON). Dayton, Ohio: IEEE Trans, on Aerospace and Electronic Systems 71-C-24-AES 283-90.

Ginsburg, A.P. (1973) Pattern recognition techniques suggested from psychological correlates of a model of the human visual system. Proc., 1973 National Aerospace Electronics Conference (NAECON), Dayton, Ohio: IEEE Trans. on Aerospace and Electronic Systems 73-CH0735-1-AES 309-16.

Ginsburg, A.P. (1978) Visual perception based on spatial filters constrained by biological data. Dissertation for Ph.D., University of Cambridge, England, (Published as AFAMRL-TR-78-129).

Ginsburg, A.P. (1980) Specifying relevant spatial information for image evaluation and display design: An explanation of how we see certain objects. Proc. of the Society for Information and Display (SID), 21/3:219-27.

Ginsburg, A.P. (1981a) Spatial filtering and vision: Implications for normal and abnormal vision. in Applications of Psychophysics to Clinical Problems (Proc. of the Symposium held in San Francisco, Calif. Oct 1978), ed. by Proenza, L., Enoch, J., Jampolski, A. The Cambridge University Press.

Ginsburg, A.P. (1981b) Proposed new vision standards for the 1980's and beyond: Contrast sensitivity. NATO/AGARD Proc. No. 310 on Aircrew Medical Standards of the Aerospace Medical Panel Specialists Meeting, Toronto, Canada, 15-19 Sep. 1980.

Ginsburg, A.P., Cannon, M.W. Jr., Nelson, M. (1980) Suprathreshold processing of complex visual stimuli: Evidence for Linearity in Contrast Perception. Science 208:619-21.

Ginsburg, A.P., Evans, D.W., Sekuler, R., Harp, S. Contrast sensitivity predicts pilots' performance in aircraft simulators (submitted for publication).

Pantle, A., Sekuler, R. (1968) Size detecting mechanisms in human vision. Science 162:1146-48.

Sekuler, R. (1974) Spatial vision. Annual Review of Psychology 25:195-232.

NEUROREDUCTIONISTIC DOGMA - A HERETICAL COUNTERVIEW[1]

William R. Uttal

Department of Psychology
University of Michigan
Ann Arbor, MI 48104

If one were asked to identify the dominant theoretical theme in perceptual science today, the answer would have to be-- neuroreductionism. Celebrated discoveries in many neurophysiological laboratories have provided a powerful heuristic for perceptual theory building. A wide variety of theories of relatively limited intended extent have been stimulated by the analogies between these neurophysiological data and perceptual phenomena. It is possible however, to classify almost all neuroreductionistic microtheories into four general classes.

1. Single Cell Feature Detector Theories

2. Fourier Channel Spatial Frequency Detector Theories

3. Lateral Interaction Theories

4. Brain Field (Electronic) Theories

The last of these four classes has long been considered to be less than viable, but the first three have dominated explanatory attempts in perception for most of the last two decades. Many psychologists have provided a wide variety of empirical and logical arguments purporting to support these neuroreductionistic theories. In this paper, however, I will play the role of the devil's (and that of an increasing number of my colleagues) advocate and argue that neuroreductionism has explanatory utility in only a limited number of perceptual situations and that this applicability is sharply limited for most interesting perceptual phenomena. My thesis will be based on a review of the presently published literature. Experimental findings will be cited as evidence against the dogma that I perceive to be implicit in neuroreductionistic theory. This essay, of course, is a reaction to Barlow's (1972) influential article in which he asserted what he believed to be the main neuroreductionistic principles in pattern detection.

In this paper I will admittedly be taking a totally one sided position; I will be largely ignoring the data that speaks for the other side. This is at least partly justified by the fact that so much of that supportive data and logic has already been presented in other places in the literature and those discussions have often been equally unbalanced in the other direction. My presentation in this essay is not, therefore, intended to be even handed, but rather to emphasize a point of view that is all too often ignored in the research literature. It is designed to emphasize the counterview that many of the popular neuroreductionistic metaphors may actually be misleading and divert attention from the true complexity of perceptual processes. For one example, I believe that many of these models involve totally inadequate isomorphic concepts that grossly misjudge the nature of the symbolic representation that actually occurs. In some cases, errors as gross as attributing a given perceptual phenomenon to a peripheral rather than a central process may be committed. As another example, the metaphors drawn from the neurophysiologiccal laboratory lead to an erroneous stress on elementalistic feature detection and divert attention from the abundance of empirical facts that point to a more wholistic processing by the human perceptual system.

Now let's consider the dogma themselves and the counterindications that should insert a cautionary note into our acceptance of a substantial portion of contemporary neuroreductionistic theory, however popular it may presently be.

Questionable Dogma Number 1

The action of single cells encode or represent complex perceptual behavior.

This very general dogma, most explicitly expressed by Barlow, (1972), is the keystone of a substantial portion of contemporary neuroreductionistic theory. In spite of the fact that neurons responding to specific trigger features of the stimulus are, ubiquitous in the nervous system, there are many logical and empirical counterarguments, that argue against the validity of such a dogma. It is clear that in many cases, we have confused the role of the peripheral nervous system as communicator with the role of the central nervous system as interpreter and psychoneural equivalent. In accepting this questionable dogma, many perceptual scientists have violated the sign-code caveat I highlighted in my earlier work (Uttal, 1973) in a particularly extreme manner. The identification of individual neuronal feature sensitivity with perceptual experience is simply not justified on logical grounds. John and Schwartz (1978) point out, the conceptual difficulty involved in the invention of a "pontifical" neuron was appreciated by Sherrington (1906) over seventy years ago. His caveat apparently fell on deaf ears.

There are several empirical arguments that can be made against the single cell hypothesis. Even Horace Barlow, the arch proponent of the neuron dogma, sees some counterindications to the single cell hypothesis. His (Barlow, 1978) inability to find any differential sensitivity to form, in a texture discrimination task, even he quite flexibly acknowledged, is a possible argument against a single cell theory of form perception.

Similarly, Timney and MacDonald (1978) raised an important question concerning feature detectors in the visual nervous system. They sought to determine whether curvature detectors per se, as opposed to multiple line detectors sensitive to the tangents of curves, were responsible for the adaptive desensitization to curved gratings by prolonged exposure to other curved gratings. They concluded that their experiments did not distinguish between the two hypothesis and also alluded to the fact that the overall structure of the pattern, and thus, "higher" levels of processing, must be involved. Pomerantz, (1978) also raises another difficulty for a simplistic single cell feature detection theory in his experimental findings that show that stimuli varying in slope alone are difficult to discriminate. While Pomerantz' main goal was to show that more complex elements than lines, such as angles, are more likely to be the basis of form perception, his results do dissociate the cellular neurophysiological data from the psychophysical in a way that is, at least, challenging to the tenets of the most simplistic of the pontifical cell theories.

The most complete body of empirical evidence counterindicating this dogma however, is the large amount of classic evidence dealing with the effect of the configuration or global pattern of a stimulus on many different phenomena. Features, unless redefined to a point of generality at which they are no longer local "properties" but rather, global aspects of the stimulus, are woefully inadequate in explaining such phenomena.

Questionable Dogma Number 2

The nervous system operates by greater and greater degrees of feature extraction and abstraction and the mapping of concepts of ever greater complexity onto the responses of an ever decreasing number of neurons.

This proposition, Barlow's (1972) second dogma, unlike the first, seems clearly to be incorrect on a strictly empirical basis. The mass of neurophysiological evidence indicates that activity initially elicited in even a single peripheral receptor neuron is magnified and distributed by neural divergence in time and space in such a way that an uncountably large number of neurons are activated in the brain by even the most localized peripheral stimulus. Rather than an increasing specificity and contraction to a small number of neurons, just the opposite seems to be happening; responses to stimuli, mediated by both the ascending reticular formation and the classic sensory pathways are generated in widely distributed regions of the brain. Nevertheless, Konorski's (1967) hierarchial theory, which is the most extreme development of this idea, pervades thinking in artificial intelligence in spite of the obvious empirical evidence to the contrary. The call for some kind of "neural economy" made by Barlow is a spurious one in a system that has many nuerons (and, perhaps more important, synapses) to spare and in which individual neurons can be involved in so many different circuits simultaneously.

Questionable Dogma Number 3

High frequency in neural responses rate encodes high stimulus certainty.

This proposition, Barlow's (1972) fifth dogma, stating that high frequency nerve action potentials (or large ones etc.) encode higher certainty of a stimulus being present, can also be criticized on empirical grounds. The weight of evidence now suggests that strong stimuli may be encoded in several parts of the nervous system either as increases or decreases from the resting level of spike action potential firing rate, as differential rates of firing in adjacent loci, or by the absolute magnitude of the response. The opponent color mechanisms in the visual system is one obvious example of this differential encoding as is the differential responsivity of binocularly sensitive neurons in the visual cortex. Such opponent or differential mechanisms probably occur in many other portions of the nervous system. The inherent isomorphism of <u>positive</u> correlations between neural and mental responses is simply not justified by current neurophysiological knowledge. A strong negative correlation may be just as "significant" as a strong positive one and ample evidence suggests that the nervous system is capable of encoding signals in just this way.

Questionable Dogma Number 4

<u>The trigger features of neurons change as the result of experience</u>.

Stryker and Sherk (1975) and Mize and Murphey (1973) have shown that, in spite of the repeated "demonstration" of experimental effects on neuronal orientation selectivity, many previously reported results of early experience may have been due to poor sampling and wishful thinking on the part of the experimenters. In their important experiment, for example, Stryker and Sherk showed that if the experiment was done with a double blind procedure and if the experimenter was very careful about using a standard spatial separation at which to gather data, then the experimental effects of early experience seemed to disappear.

Questionable Dogma Number 5

<u>Neuroanatomically distinct channels selectively sensitive to spatial frequency exist in the visual nervous system</u>.

This, of course, is the prototypical dogma of the Fourier analysis type of neuroreductionistic theories. As I have noted, this model can hold in this strong form only if two very specific properties of the channels can be validated; the channels must be characterized by narrow tuning and interchannel independence. Without these two properties, as Towe (1975) has noted, there is no way to distinguish between this specific mechanism and a host of others that can also represent spatial frequencies because of the very general nature of the Fourier postulate. Almost any nervous system that is made up of neurons arranged in some sort of spatial array or composed of a field of various size receptive fields will exhibit some properties that are analogous to the Fourier channels. Thus, the rigorous proof of a Fourier-type theory demands direct neurophysiological validation of

narrow tuning and channel independence, validations which I have
already indicated are not yet available.

Psychophysical evidence that the molar behavior is "consistent"
with these properties is not a proof of the model. However, any
psychophysical evidence that shows that the molar performance is not
consistent with the channel hypothesis would be compelling. And,
indeed, there is a substantial abundance of information that so argues
against the two prerequisite Fourier premises. For example, Stromeyer
and Klein (1975) have presented evidence that seems to support the
idea that the channels if they exist must be very broadly (and not
narrowly) tuned. Tolhurst (1972); Strecher, Sigel, and Lange (1973);
Henning, Hertz, and Broadbent (1975); Nachmias and Weber (1975); and
Tolhurst and Barfield (1978) on the other hand, have presented equally
compelling evidence that the channels, if they exist, must be
interacting. If these findings are valid, then Towe's criticism of
indistinguishability holds and no Fourier channels could ever be
indisputatively identified in an anatomical sense.

A wide variety of other inconsistent psychophysical data is also
available to counterindicate the neuroanatomical-channel postulate of
the most extreme kind of Fourier theory. The band width of any
channels that might exist seem to vary quite a bit depending upon the
method used (See Graham, in press, for a discussion of this point) and
the theory typically fails (like most theories) to provide a good fit
for the psychophysical data at the high and low frequency limits of
the range. Furthermore, Coffin (1978) has shown that the Fourier
theory fails very badly when it is confronted with a task that does
not involve stimuli already represented as spatial frequency patterns.
If stimuli such as block letters are used and the Fourier analysis is
only done ex post facto, he reports that the confusions among such
letters obtained in psychophysical experiments are not well predicted
by the Fourier components of the stimuli. Indeed, the predictions of a
more conventional feature similarity model fit the data much better. A
further counterindication is that Growney (1978) has shown that the
Fourier model does equally poorly in explaining the metacontrast
phenomenon even when the stimuli were explicitly spatial frequency
patterns.

Another counterindication to the Fourier channel hypothesis can be
found in the work of Greenwood (1973). In an interesting study he
compared the contrast sensitivity of patterns that were composed

either of lines (gratings) or of grating dot patterns in which the elements were arranged with the same spatial frequencies as the gratings. The results of this experiment showed that the gratings were detected at contrasts less than 40% of that required to detect the grainy patterns even when matched for brightness. Obviously the visual system does not depend on Fourier channels alone, if at all; other dimensions of the stimulus such as the local microstructure are also very important. Furthermore, in a backward masking experiment (Zamansky and Corwin, 1976) in which scattered letter fragments were used to mask letters, fragments of the same stroke width as the letters were not the best maskers as would be predicted by a spatial frequency-type hypothesis.

Clearly the Fourier model seems to work best in the context of grating-like stimuli and, even then, not universally as proposed by its proponents. Another phenomenon that is very difficult for the radical theory of Fourier channels to handle is Georgeson's (1976) observation that the afterimages produced following persistent viewing of a grating have an apparent spatial frequency 1.5 octaves greater than the adapting one. Despite a somewhat tortuous bit of neural backsliding on his part, invoking hypothetical lateral inhibition between the hypothetical channels, Georgeson's data remains, for some of us, another discrepancy in the conceptual structure surrounding Fourier channel theory.

There are many other logical and mathematical arguments against the strong neuroanatomical Fourier channel idea. Poggio (1979) has noted the following:

1. "It is impossible to represent Fourier coefficients by means of cells with a receptive field that is spatially localized."

2. "Vision is generally nonlinear and the role of Fourier analysis in nonlinear systems is equivocal."

3. "The number of channels (12) reported does not seem to be large enough."

4. "No convincing demonstration has yet been made that shows that phase information (a necessity for perceptual reconstruction) is encoded and yet pictures with identical power spectra may look quite different."

5. "The notion that a simple cell with a bar-shaped receptive field acts as a bar detector is simplistic."

Questionable Dogma Number 6

Metacontrast is a result of peripheral lateral inhibitory interactions that diminish the strength of the conducted neural signals.

This dogma is perhaps one of the most pervasive in modern perceptual psychobiology. Surprisingly it is also the one for which the most abundant body of counterindications exist. For over fifteen years there has been an accumulation of solid findings that argue against the lateral interaction theories of metacontrast that have been proposed by a number of investigators. Unfortunately these data are largely ignored in the enthusiastic debates concerning which version of the lateral interaction model best represents the psychophysical data. Yet, over a decade and a half ago Fehrer and Rabb (1962) raised a very important caveat when they noted that the reaction time to a metacontrasted stimulus (which is presumably perceptually less intense) did not differ from that measured to the same, but uncontrasted, stimulus. Since the link between brightness and reaction time is so tight, this should have been a prominent warning that the diminution in the metacontrasted percept should not be equated with a diminution in the afferent signal strength. The phenomenon is obviously much more complex than the simple lateral inhibition model suggests if its impact on reaction time can remain constant even though the "brightness" varies to virtually zero.

Several more recent studies have also shown that the process is dramatically dependent upon higher levels of cognitive processing than are dealt with by the simple lateral interaction model. Mayzner and Tresselt (1970), for example, in studying the powerful form of metacontrast that they called sequential blanking, should have had their suspicions raised about the lateral interaction model when they discovered that a sequential series of letters blanked each other only when the letter sequence did not form a word. My own study of metacontrast (Uttal 1970) was prima facie evidence that lateral inhibition was an inadequate explanation. In that case the contrast effect seemed to depend upon the contrasted and the contrasting parts of the stimuli sharing a common form. Since the contrast effects require that a prior form recognition and comparison be made, it is hard to appreciate how the metacontrast suppression could be as peripheral and passive as suggested by the lateral interaction model. Thus, unless one is willing to put such complex cognitive functions as form recognition in the retina, the model fails rather badly.

Over a decade ago it was known that the suppressed response to a metacontrasted stimulus could be recovered by subsequently suppressing the metacontrasting pattern. This was originally shown by Robinson (1966) and then further supported by Dember and Purcell (1967). This temporal "disinhibition," as it was called by these workers, is not analogous to the spatial disinhibition observed in the Limulus eye by Hartline and his collaborators. In the metacontrast paradigm the time difference between the original stimulus and the disinhibitor is so great that a lateral interaction theory would have to involve some kind of storage mechanism that is not a part of that simple passive model. Such a storage mechanism immeasurably complicates the lateral interaction explanation, to the point of converting it into a more constructionistic type explanation.

Furthermore, the suppression of the contrasting stimulus by a disinhibitor is not a requirement to restore the original stimulus to visibility. Such a trick can be performed simply by changing one's research methodology! Schiller and Smith (1966) and Pollack (1972), using forced choice experimental designs, have both shown that the metacontrasted information, even though suppressed to invisibility, could be recovered simply by altering the subject's criterion. This is also strong evidence that the neural signal information is not passively inhibited, but rather is actively suppressed in such a way that it loses perceptual significance without signal degradation.

Hernandez and Lefton (1977) have further shown that there is a profound shift in the criterion used by the subject at exactly the place where the slope reverses in the typical U-shaped metacontrast curve. This result also suggests that there is a strong cognitive or interpretive component to the nonmonotonic metacontrast function that is totally ignored by the lateral interaction hypothesis.

Additional counterindications to this dogma are available. Hogben and DiLollo (in press) have found substantial practice effects in the metacontrast paradigm as well as an absence of any effect of the number of contours that are interacting. The net impact of their results is to reinforce the view that the metacontrast function is not the result of simple spatial interactions; the geometrical factor (number of inhibitors) that should be effective is not, and the experimental factor (practice) that should not be effective actually is.

A further substantial counterindication to the peripheral lateral interactions hypothesis of metacontrast can be found in experiments that have varied the apparent depth between the inducing field and the target. It is now well established (Fox and Lehmkuhle, 1978) that the metacontrast effect diminishes as the difference in depth between the two increases. Since the monocular retinal images still maintain the same spatial adjacency (and thus potential for lateral interaction), but the effect diminishes, once again it seems unlikely that the lateral interaction hypothesis is valid.

The general conclusion to be drawn from all of these data is that there are sufficient cognitive, symbolic, and relational factors involved to raise a serious question about the isomorphic, passive, lateral interaction explanation of the metacontrast phenomenon. To this deviant psychophysical data we can couple the fact that there are virtually no well substantiated physiological findings that unequivocally show metacontrast-like effects in peripheral nervous tissue. Even Bruce Bridgeman, a strongly committed neural interaction theorist, was unable to find (Bridgeman, 1975) any backward masking in the cat's optic tract or lateral geniculate nucleus that corresponded to metacontrast, even though he did find some attenuation of a delayed burst of activity in the cortex that he felt might be comparable in some ways to this psychological experience.

Finally I should note that the mathematical assumptions of the various lateral interaction theories are separable from the physiological ones and even given the fact that it is possible to fit the psychophysical data relatively well with such equations does not necessarily support the neurophysiological postulates of the model.

Questionable Dogma Number 7

   Simultaneous contrast is a result of the same kind of lateral
inhibitory interaction mechanisms that seem to adequately explain the
Mach band and the Hermann grid.

   It has been virtually axiomatic among visual scientists that the
Mach band and simultaneous contrast are expressions of one and the
same process--reciprocal lateral inhibitory interactions. Such a
distinguished visual scientist as Floyd Ratliff (1965) has suggested
that Mach bands and simultaneous contrast "share some underlying
physiological mechanisms in common." However, a closer examination of
the similarities and differences between these two sets of phenomena
suggests that they are actually very different. Simultaneous contrast
is a global effect that occurs over the entire enclosed region
regardless of distance from the edge, while the Mach bands are highly
localized effects, highly sensitive to the distances between the
interacting regions. Indeed, our very successful theoretical
understanding of the Mach band effect depends entirely upon this
strong relationship between the strength of the interaction and the
distance between the interacting objects.

   Another discrepancy concerns the locus of the two effects. I have
previously presented a demonstration (Uttal, 1973) that suggests that
the simultaneous contrast effect is quite central and can be produced
by dichoptic display in which neither eye alone could individually
give rise to the effect. Julesz (1971) makes this same point with a
random dot stereogram on page 323 of his delightful book on the
cyclopean eye. On the other hand, Werner Koppitz (1957) has shown
that the Mach band does not appear when the parts of the stimulus are
shown separately to each eye. Koppitz' conclusion was reinforced by
another stereoscopic demonstration of my own (Uttal, 1973) in which it
was shown that the Hermann grid illusion could not be produced by
dochoptic fusion.

   Further support for the idea that simplistic lateral inhibitory
interaction ideas do not adequately explain brightness contrast can be
found in a very interesting paper by Kaniza (1975). In this paper he
presents a wide variety of brightness contrast displays in which the
form or configuration of the stimulus plays a critical role in
defining the effect above and beyond consideration of simple luminance
and geometry.

Additional counterindications are available. Gogle and Mershon (1969), Mershon and Gogle (1970), Mershon (1972), and Gilchrist (1977) have all shown that the simultaneous contrast effect as well as the metacontrast phenomenon, depends upon apparent adjacency in depth and that, if the induced and inducing fields are separated (in depth), the effect is substantially reduced. Other studies have shown strong cognitive components in the simultaneous contrast paradigm that are not consistent with the lateral interaction model. Festinger, Coren, and Rivers (1970) showed that the degree of effort with which the subject attended to the pattern could result in a modulation of the magnitude of the effect. The entire literature on lightness contrast, if considered to be parallel with the kind of brightness contrast effect I have been discussing, constitutes another major counterindication to the passive lateral inhibitory model by involving active relational processes. It has been known for over a hundred years that the lightness of an object depends upon our knowledge of what it is or that it is continuous with an object that may be of a widely different luminance. Recent research has supported this classic observation by manipulating the apparent spatial position (Beck, 1965) or the figure-ground relations (Coren, 1969).

Finally, specific searches (e.g., DeValois and Pease, 1971) for a neural correlate of the simultaneous contrast have been unsuccessful in the peripheral nervous system of mammals even though the localized interactions corresponding to the Mach band have been observed.

Questionable Dogma Number 8

Prolonged viewing leads to figural after-effects that are the direct result of fatigued neural components.

There is perhaps no more hoary dogma in psychology than that which proposes that figural aftereffects such as the waterfall illusion or the McCullough effect are the result of some kind of neuronal fatigue. In particular, a differential adaptation in an otherwise balanced bivalent or opponent system has often been proposed as the locus of such aftereffects. In recent years, however, some extraordinarily powerful counterindications have been forthcoming that also challenge this dogma. These counterindications typically involve the duration of the "adapting" or fatigue effect. Specifically we have known, or, perhaps, have overlooked the knowledge, for nearly a decade, since the work of Stromeyer and Mansfield (1970), that aftereffects produced by prolonged viewing of spirals or stripes can last for over six weeks! More recently, this prolonged persistence of aftereffects has been shown to last even longer. Jones and Holding (1975) demonstrated that the McCullough effect could last up to six months as long as the subject was not exposed to the specific test stimuli during that time! Once tested, however, the effect rapidly diminished following the same time course as if the tests had commenced at the end of the adaptation period. Others who have supported this extraordinarily long persistance of the McCullough effect include Riggs, White, and Eimas (1974), Skowbo, Gentry, Timney, and Morant (1974), MacKay and MacKay (1975), Heggelund and Hohmann (1976), and Thompson and Movshon (1978), and White (1978). The extraordinary durations of these aftereffects are hardly consistent either with what is known electrophysiologically about the recovery of function of individual neurons, or what might plausibly be inferred about the recovery of complex networks from usage.

Further support that the perceptual effects are underlain by processes much more complicated than simple fatigue has been presented by Sharpe and Tees (1978) who have demonstrated a remarkable lack of influence on the figural aftereffects when stimuli are blurred or interrupted in time! Additionally, they were unable to demonstrate any difference between early and later conditioning trials--an effect that would have been predicted by a simple fatigue model.

Another counterindication suggesting that the McCullough effect is not mediated by tired neurons is Jenkins and Ross's (1977) demonstration that the effect switches in and out as a function of the subjective state when the figure is a reversible one. Their reversible figure could be seen either as a set of triangles or a set of squares. If the McCullough effect was induced by inspection of a green vertical-red horizontal grating, the phenomenon would occur when the triangles were the perceived state of organization of the figure. However, when the perceived state was the set of concentric squares the effect would disappear. Since the subjective organization and interpretation of the inspection figure plays such an important role, it is unclear how any simple theory of isomorphic neural fatigue could account for this finding.

Obviously, simple fatigue models do not perform well in explaining aftereffects. One has to look elsewhere, most probably at higher order

206

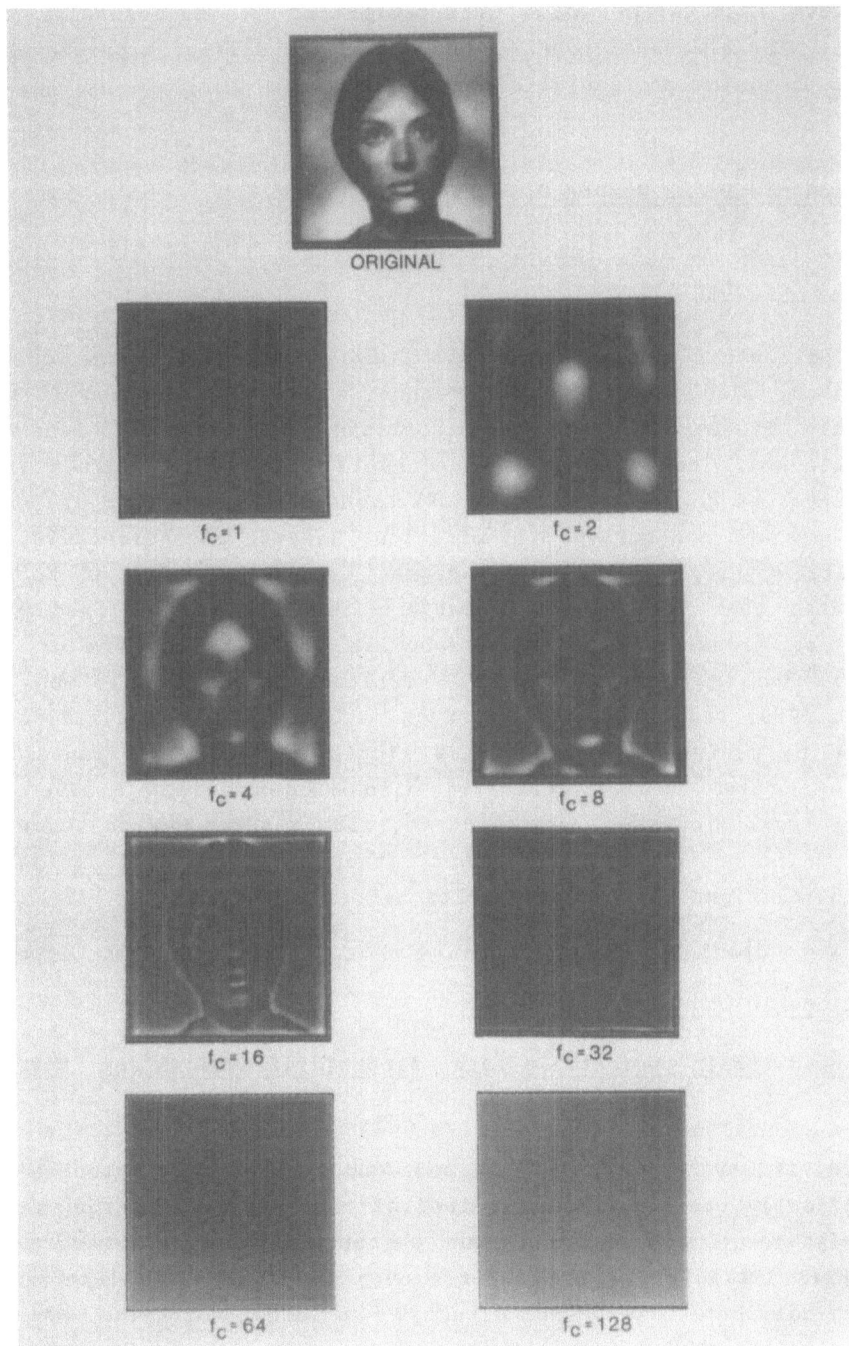

Figure 3: A hierarchy of filtered images created from channel filters based on biological data. The original portrait is filtered into eight channels of spatial information having a bandwidth of two octaves and center frequencies one octave apart (from Ginsburg 1978, 1980).

cognitive effects in which interpretations of the stimulus become salient, for a more appropriate answer to the difficult problem raised by the phenomena of figural aftereffects.

Questionable Dogma Number 9

Stabalized retinal images lead to percept fading due to fatigued retinal components.

A perceptual dogma which is closely related to the one just described, also possibly false, concerns the oft reported fading of retinal images that are stabalized by contact lens-mirror arrangements. The explanatory hypothesis that is most often invoked also involves the decreased responsiveness of neurons that are exposed to a constant contourless illumination. However a considerable amount of data counterindicates this dogma, too. One body of knowledge indicates that the perceived pattern in such a stabalized viewing situation fades in meaningful subunits (such as corners or sides) rather than in random blotches. (Pritchard, Heron and Hebb, 1960; Evans, 1965; and Davies, 1973). Such a context-dependent fading could hardly be mediated entirely by a simple peripheral mechanism. This finding, quite to the contrary, also suggests the involvement of higher level processes. The point is, fading seems not to be due to a loss of the peripheral signal, but rather to a "designification" of a continuing signal by a more complex central mechanism.

Questionable Dogma Number 10

Evoked brain potentials are direct reflections of perceptual states.

Few, if any, students of the compound evoked brain potentials have specifically spelled out the nature of the relationship they perceive to exist between these signals and perceptual responses. Nevertheless, there is implicit in this entire research program the idea that the electrical brain response directly reflects something about the psychological state that a similar electrical response from the liver, for example, would not. In other words, the prevailing implicit dogma with regard to these evoked potentials is that they are psychoneural equivalents or codes and not simply systemic signs of a generalized brain activity.

This dogma is especially frail and would fall victim to any findings that show disassociation of the perceptual response and the evoked potential. And, indeed, several such disassociative studies have now appeared. Recently, an extraordinary case of dissociation involving evoked brain potentials was reported by Bodis-Wollner, Atkin, Raab, and Wolkstein (1977). They carried out a more or less conventional evoked potential study on a very unconventional subject. The subject was a six-year-old boy who had massive destruction of the occipital lobe of his brain. The damage, however, was restricted in a way that spared the primary visual projection regions of Area 17. In spite of this damage, a normal evoked potential to a checkerboard stimulus pattern could be obtained. Their data was especially remarkable in that the evoked potential was normal in <u>all</u> regards. Not only where the initial transients (generally associated with activity in the primary projection regions) present, but the later potentials (200 msec following the stimulus) that have been associated with cognitive manipulation of the afferent information were also normal. Psychophysically, however, the child was totally blind in terms of the results of a battery of behavioral tests. Thus, even though these evoked signals did indicate the presence of neural activity in the afferent pathway, the microvoltages seem to be totally unrelated to any perceptual experience. This is prima facie evidence that the evoked potential is not a direct indication of the percept, but rather, probably only reflects the activity of the afferent communication system. Though this distinction may seem trivial, the dogma that the evoked potential is a psychoneural equivalent of perception has already been developed into the practice of clinical testing of vision or audition. Results such as that of Bodis-Wollner and his colleagues make this a highly doubtful procedure in doing anything other than testing the integrity of the afferent pathway.

Questionable Dogma Number 11

<u>Many geometric illusions and other related perceptual phenomena are due to peripheral distortions of the neural representation of the stimulus.</u>

Among the most mysterious, and yet familiar, phenomena of visual perception are the geometric illusions. Although theories of visual illusions are ubiquitous, we have but the vaguest idea of the mechanisms that underly these intriguing and fascinating illusions. A masterful review of the state of theories of geometrical illusions has been presented by J. O. Robinson (1972) in what is probably the best general book on illusions yet available. Robinson describes a number of neuroreductionistic theories, most notably those of Eriksson (1970), Motokawa (1950), Motokawa and Akita (1957), Chiang (1968), Kohler and Wallach (1944) and Ganz (1966 a and b). All of these theories share a common feature – they all invoke a spatial interaction somewhere in the (presumably) more peripheral portions of the nervous system among the parts of the stimulus pattern as the necessary and sufficient cause of this class of illusions. The theories invoke isomorphic distortions rather than interpretive ones and see this distortion remaining a constant influence as the coded signals ascend and are transformed to the perceptual experience itself.

It is difficult to designate generalized experimental evidence that attacks all of these theories collectively. There are general logical or conceptual arguments that can be marshalled against them as a group, however. First, the implicit and radical isomorphism that is a part of all of these theories ignores the abundant evidence that suggests that isomorphism is a treacherous foundation upon which to build a theory. Second, there is a total absence of first-class physiological evidence to support the idea that the perceived spatial distortions are present in the peripheral neural interactions. In spite of the many traditional models and theories of these neural "induction" effects (see, for example, Obanai's, 1977 summary of such theories), there is yet to be any validation of these processes found in electrophysiological laboratory.

There is, however, specific evidence that each of the neuroreductionistic theories of geometrical illusions fails in one way or another. Robinson adequately describes the weakness of the experimental methods utilized or the failure to properly predict outside of a very narrow range of phenomena that undermine the assumptions of many of these theories. He notes that, Ganz's lateral interaction theory comes to grief by incorrectly predicting the relations between simultaneous and successive effects as well as in its underprediction of the spatial extent of the illusions. Chiang's diffraction theory clashes with phenomena other than the ones with which he specifically dealt. Motokawa's electric-photic interaction experiments have proven difficult to replicate. Others like Eriksson's field theory seem to be no more than the invocation of a physical analogy as a metaphorical description rather than as a true reductionistic explanation. Other studies that show that the "inducing" effects of one stimulus on another are reduced if the two are at different depths, also present difficulties to lateral interaction theories since the x, y dimensions are all but identical in a scene with and without depth. For example, Gogel and Newton (1975) have shown a marked diminishment in the rod and frame illusion when the rod and frame are presented at different depths.

Another general difficulty for any theory invoking spatial interactions between parts of the stimulus is that the stimulus components that must supposedly interact need not be physically present to effectively "distort" some other feature of the stimulus scene—they need only be suggested. Coren (1970), for example, has shown that the Muller-Lyer illusion occurs even though the stimulus lines are reduced to dot patterns. In other experiments, for example, those of Goldstein and Weintraub (1973) and Kaniza, (1976) illusory contours can also be shown to possess the same perceptual effectiveness as real contours even though they are present, so to speak, only in the eye of the beholder. Similarly the tilt illusion can be produced by subjective contours according to Smith and Over, (1977). Spatial interactions between the fields of electronic influence of nonexistent component parts become a nonsensical concept, and if suggestion and symbolic construction play such a great role in this case, why not also in the case in which the contour happens to be physically present?

A similar counterindication to a peripheral explanation of movement illusion can be found in the work of Robinson and Moulton (1978). They showed that the induced apparent movement exerted by a moving dot on a stable one occurred after a lag of .33 sec! They concluded that no simple, peripheral isomorphic explanation would explain such a phenomenon. Coren and Girgus' (1978) comprehensive and insightful book on visual illusions also invokes other "structural" factors that might contribute to many of the classic visual illusions. However they too (see p. 116) appear to think that there are many perplexities in applying such ideas to explaining these illusions.

In sum, geometrical illusions, while clearly responsive to the spatial properties of the stimulus, do not so clearly involve any kind of spatial interaction between isomorphic neural representation.

Questionable Dogma Number 12

   Humans see by analyzing stimulus pattern into features.

   There may be no more misleading dogma among those I have considered
so far than this one. The entire single cell neuroreductionstic
approach feeds back and forth onto the idea that features (to which
neurons are supposed to be sensitive) are more important in human
perception than the global organization of the stimulus. It is as if
the great historical debate between elementalism and wholism never
occurred--so great is the commitment to elementalism in current
perceptual theory. Perceptual theorists have swept up, more or less
uncritically, the concepts of local feature detection, feature
recognition, and feature analysis. This enthusiasm is obviously
stimulated by the exciting progress in the neurophysiology laboratory
as well as in computer science; the microelectrode technology and the
logic implicit in present day programming arts both emphasize such an
elementalistic approach.

   In spite of this wholesale commitment to a feature oriented
elementalism in current theory, it seems clear that there has been
ample evidence for many years in the visual literature that humans
perceive not by features but mainly by the global configuration of the
stimulus. However disappointing the theoretical contributions of
Gestalt psychology were, it is impossible to deny the many instances
to which they called attention that strongly argued for a wholistic,
as opposed to a featuristic, processing of visual stimuli by men.

   The classic fractured figures of Leeper(1935) and Kolers' (1970)
famous chairs clearly indicate that the particular angles, curves, and
lines from which a pattern may be constructed are less important than
the overall configuration into which the "features" are arranged.
Unless one wishes to play a word game and call the global organization
just another feature, obviously these demonstrations are strong
counterindications to the idea that man sees by processing features.
My own work on an autocorrelational analysis of form detection (Uttal,
1975) also makes this same argument for a wholistic approach as does
the work on the superiority of line detection that occurs when the
line is part of a configuration as demonstrated by Weisstein and
Harris (1974). Other workers have also shown strong context effects in
perception (e.g., Pomerantz, Sager, and Stoever, 1977) and Prinzmetal
and Banks (1977). All such experiments speak against the basic

premises of the radical feature analytical model of perception.

Questionable Dogma Number 13

It is possible to tell from behavioral studies whether any particular coding scheme is used to represent stimulus quantity or, for that matter, what code is used in the nervous system to encode any perceptual dimension.

Many perceptual theorists seem to have forgotten that this sort of internal analysis by means of behavior is simply not possible for the surest mathematical and logical reasons. Plausible mechanisms may be suggested, implausible ones rejected, and demonstrations made of the processes that must be executed but, in the final analysis, no behavioral test can say anything definitive about internal neural codes, any of which can equally effectively represent any concept or idea. The many behavioral studies that purport to demonstrate neuronal receptive fields, channels, feature sensitivity, or orientation detectors, or counting or timing internal codes in the human nervous system are all logically inadequate in the light of this important constraint.

Questionable Dogma Number 14

Apparent motion produces an internal neural response that is identical to real motion.

Kolers (1964) has shown that the path of an apparently moving object is not able to inhibit a small stimulus in the same way that the path of a real movement does. Thus, apparent motion, like geometric illusion, must involve higher level interpretive processes than the isomorphic ones invoked by a simplistic neuroreductionism. An important general implication of this specific result is, therefore, that no isomorphic neural signal need be associated with a perceptual response; apparently symbolic interpretations can suffice. This is a strong argument for a constructionalistic, if not rationalistic, model of this aspect of human perception.

Questionable Dogma Number 15

Binocular rivalry is due to a reciprocal lateral inhibition between the afferent signals from the two eyes.

In a comprehensive review of the problem of binocular rivalry Walker (1978) masses a substantial amount of evidence that argues against a lateral inhibition idea between the two eyes as an explanation of the rivalry phenomenon. In my opinion, his review more than adequately demolishes this traditional dogma.

Questionable Dogma Number 16

Stereopsis is the result of either suppression or fusion of the two monocular images.

Although it is difficult to find specific empirical evidence to counterindicate this persistent dogma that stereopsis is due to either suppression or fusion, it seems to me that both psychophysical and neurophysiological data counterindicate both sides of this false debate. There are, furthermore, logical reasons to assume that stereopsis is not a result of either the suppression of one (or the other) monocular image or the fusion of the two for the simple reason that the stereopscopic image is neither the sum, nor the difference of the two images, nor the manifestation of either unsuppressed one alone. Rather, the three dimensional experience is a totally new psychological construct that takes the place of the two monocular impressions. There is no way that the two-dimensional images impinging on the two retinae can themselves be manipulated spatially or geometrically to produce a three-dimensional image. Instead, the transformation has to be one in which the two images either provide informational cues that must be evaluated by some interpretive process or recoded into a new three-dimensional language. The search for a residual two-dimensional percept that no longer exists (and may never have existed) is a search for a chimera. It, too, was nothing other than the most reasonable perceptual construction of the available two-dimensional cues. However, when sufficient information is available to construct the third dimension, the stereo percept is the one constructed and the two-dimensional one simply does not exist.

A specific empirical counterindication to the suppression hypothesis can be found in Blake and Camisa (1978). These workers showed that there was no reduction in visual sensitivity to either eye during stable stereoscopic (cyclopean) vision comparable to that found in retinal rivalry. Thus, the vision from each eye is not suppressed during binocular vision, a suppression hypothesis of stereopsis is patently untenable.

Questionable Dogma Number 17

The perceived color of a stimulus is solely defined by the relative amount of activity in three cone receptor systems or by the stimulus spectrum.

Although there is considerable evidence that the wavelength spectrum emitted by or reflected from an isolated object in space is mainly encoded by the relative degrees of activity induced in the three cone systems, it is clear that the spectral properties of the stimulus are not, in a more general sense, the only determinant of perceived color. Color contrast effects, which are not well explained by lateral interaction models, are supplementary influences. A color constancy effect, in which the color of an object may depend in complex ways upon the ability of the human to compute the chromatic reflectivity, or lightness, of the object represents another counterindication to simple spectrum -- percept veridicality.

Perhaps the most compelling counterargument to this particular dogma comes from the work of Edwin Land (see, particularly, Land, 1977) in which he showed that red, green, or blue experiences can all be produced by exactly the same flux of spectral energies. The chromatic experience in this case depends upon the relationships among the various objects in the stimulus. These effects, though not yet completely understood, seem to be directly "computed" from the respective reflectances of the objects and do not depend, as often suggested, upon any knowledge of what the color "should" be. However, the known color of an object (e.g., an egg or a piece of coal) does affect perceived color or lightness in other situations and, therefore, also represents an additional counterindication that the spectral properties of the stimulus are the sole determinants of perceived chromaticity.

The point here, is that the detailed knowledge that we have of receptor functions and trichromatic color addition should not be overgeneralized as the sole determinant of all aspects of color perception. The evidence is overwhelming that those trichromatic data represent the outcome of a highly specific and abstract experimental situation.

Questionable Dogma Number 18

There is an isomorphic relation between certain dimensions of the stimulus and certain dimensions of the perceptual response.

Dogma 17, can, in fact, be generalized for most perceptual dimensions. A major conclusion from much of the perceptual research that has been done on illusions, contrast, and constancy is that there is no direct deterministic link between any particular aspect of a stimulus and the associated percept even though they may often be closely correlated. Whether it be in the domain of space, time, quality or quantity, we see innumerable examples of stimulus-percept nonveridicality. In some cases the correlation between the stimulus properties and the perceptual experience may be exactly zero; a single stimulus property can be interpreted in many different ways depending upon the context and significance of other parts of the stimulus scene. We apparently have been devastatingly misled by the design of the typical laboratory experiment to misassume that the neat unidimensional relationship observed between a single aspect of the stimulus and a single aspect of the percept holds in general. While systematic unidimensional relationships between stimuli and percepts do exist, they usually occur in contexts in which all other factors have been held constant or removed from the stimulus scene. In situations that are more ecologically valid, the one-for-one relationship does not always hold. As a result we see such discrepancies as size, lightness or color constancy in which the percept deviates substantially from that defined by the stimulus. No stimulus dimension, in short, uniquely defines a perceptual dimension. Each stimulus dimension or aspect contributes to the aggregate of cues from which the perceptual experience best solving the problem posed by those cues is constructed.

In this essay I have challenged some of the deeply held and popular dogma of current perceptual theory. There are many other microtheories of the sort that I have described here that might also justifiably be classified as questionable, though widely accepted, dogma. It is not necessary, however, to push this discussion further. I hope by now that I have made the general point that many beliefs in contemporary perceptual neuroreductionism are highly suspect. Although each of the arguments that I have presented here could be countered by an abundance of other more positive data, the simple fact is that there does exist a body of empirical counterindications and logical counterarguments that strongly suggests that we may have prematurely adopted a more accepting point of view about neuroreductionism than is currently justified. I am suggesting here that the complexity of the neural mechanisms underlying most perceptual processes has been grossly underestimated in current thought. Unsupported, but widely accepted, metaphors and analogies may have led us to profoundly overestimate both the level of understanding of perceptual phenomena that we have achieved, and what it will be possible to achieve within the limits of the neuroreductionistic approach.

Now that this point has (hopefully) been made, it would be well to look at an important general problem raised by this criticism. If I am correct so far, how then, do we account for the propensity on the part of so many very bright people to so uncritically accept these generally unsubstantiated dogma? The answer to this question is psychological in another sense of that word.

The factors contributing to the contemporary overcommitment to empiricistic neuroreductionism can be briefly tabulated as follows:

1. First and foremost, or course, is that the problem of how neural mechanisms represent mental processes is important and is yet unsolved. Nevertheless, people want answers. It is the essence of the human spirit to search out answers and sometimes to invent them when they are not yet actually available.

2. Throughout history the response to the great psychobiological perplexity has been to invoke the contemporary technology as the model of mind in general or perception in particular. Thus, today's most powerful heuristic technology for perceptual theory appears to be single cell neurophysiology. Since neurons do encode thoughts in some way, our knowledge of the afferent and efferent communication system has become a model of central processes in exactly the same way that telephones and computers did in their day.

3. We tend to confuse superficial process analogies with true structural homologies. Lateral interactions between contours and simultaneous contrast are not homologs, only partial analogs of each other.

4. We tend to misunderstand the intended range of application

of an idea. There is considerable confusion in the literature concerning whether the Fourier channel theory is only an abstract mathematical description of the stimulus or internal representation, or does it assert the actual existence of anatomically distinct spatial frequency channels in the brain?

5. There are, in fact, few direct neurophysiological and psychophysical comparisons of truly comparable experimental situations. The linkage being so poor, we have a tendency to mistake analogies for homologies. We lack appreciation of the fundamental message of coding theory, namely, that any code can represent any message.

6. Conversely, we tend to underappreciate the possible role of symbolic representation in perception -- symbolic representation that is also mediated by neurons, but neurons that are interconnected in a network of such complexity that we may never be able to link the neural logic with the mental processes for reasons of simple logistics and computatability.

7. We have a tendency towards overly simplistic peripheral explanations of complex central processes (e.g., the totally inadequate models of simultaneous contrast and metacontrast).

8. There is a semantic confusion in the use of key words (e.g., thresholds, learning, masking). Neural and psychological processes are sometimes described with the same vocabulary in a way that can only be described as a pun.

9. There is a poor definition of the perceptual constructs that people are trying to model. Only recently, for example, has signal detection theory been specific concerning the relative magnitudes of the discriminative and criterion contributions to thresholds, masking, etc.

10. We overlook the great differences in time constants between proposed neural explanations and the supposedly similar psychological processes.

11. We misidentify stimulus dimensions in what are actually nonanalogous situations.

12. We ignore basic perceptual data.

13. There is a profound absence of a good classification system of perceptual processes. Phenomenologically organized texts obscure similarities and falsely emphasize analogies.

14. Neuroreductionstic theories in perception are usually not presented as well defined propositions that can be tested for logical clarity or consistency. Very often the analogies and models are very loosely formulated.

As negative as this tabulation is in tone, there is another side to it. Even though the errors of logic and concept that I have alluded to here may all have been committed, it may have been a necessary stage through which any science must pass. However sincerely we may disagree with the neuroreductionistic philosophy, we all learned from it. Many other perceptual scientists are now becoming aware of the limits of neuroreductionism and there is an increasingly critical (in the good sense) body of literature appearing. Furthermore, there may have been no alternative. I acknowledge the incredible complexity of the perceptual nervous system and the frailty of our intellectual and instrumental armamentarium. The important fact is, however, that we have to pass on and one of the most important tasks ahead is to define the point beyond which we must admit that neuroreductionism fails us. All agree with this statement. The major problem in this area is to identify that point of demarcation between the plausibly reductionistic and the practically irreducible. I propose the following taxonomy as a possible answer to the location of the limits of neuroreductionistic analysis.

Footnote

[1]This essay is adapted from a Chapter in a forthcoming book entitled A Taxonomy of Visual Processes and is used with the permission of the publisher, Lawrence Erlbaum Associates, Inc.

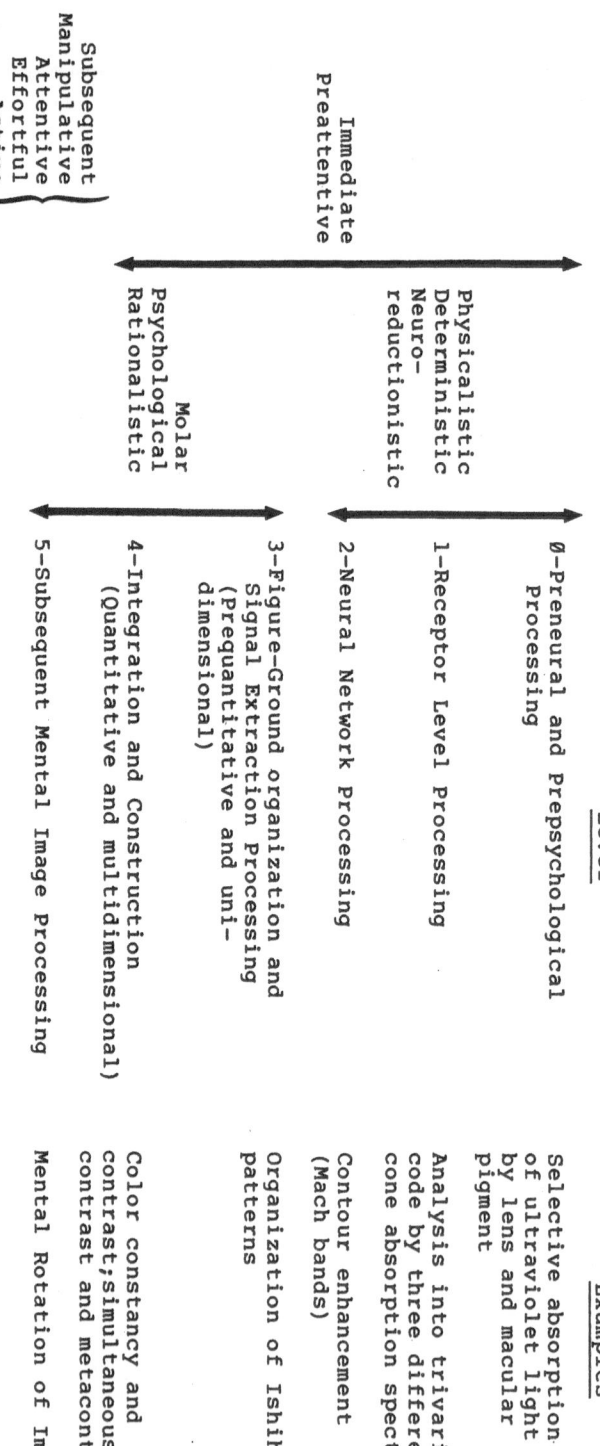

219

Table 1

Level	Examples
Physicalistic Deterministic Neuro-reductionistic	
0-Preneural and Prepsychological Processing	Selective absorption of ultraviolet light by lens and macular pigment
1-Receptor Level Processing	Analysis into trivariant code by three different cone absorption spectra
2-Neural Network Processing	Contour enhancement (Mach bands)
Molar Psychological Rationalistic	
3-Figure-Ground organization and Signal Extraction Processing (Prequantitative and uni-dimensional)	Organization of Ishihara patterns
4-Integration and Construction (Quantitative and multidimensional)	Color constancy and contrast;simultaneous contrast and metacontrast
5-Subsequent Mental Image Processing	Mental Rotation of Image

Immediate Preattentive — levels 0, 1, 2

Subsequent { Manipulative, Attentive, Effortful, Active } — levels 3, 4, 5

References

Barlow, H.B. 1972. Single units and sensation: A neuron doctrine for perceptual psychology. Perception 1, 371–394.

Barlow, H.B. 1978. The efficiency of detecting changes in random dot patterns. Vision Research 18, 637–650.

Beck, J. 1965. Apparent spatial position and the perception of lightness. Journal of Experimental Psychology 69, 170–179.

Blake, R., and Camisa, J. 1978. Is binocular vision always monocular? Science 200, 1497–1499.

Bodis-Wollner, I., Atkin, A., Raab, E., and Wolkstein, M. 1977. Visual association cortex and vision in man: Pattern-evoked occipital potentials in a blind boy. Science 198, 629–631.

Bridgeman, B. 1975. Correlates of metacontrast in single cells of the cat visual system. Visual Research 15, 91–99.

Chiang, C. 1968. A new theory to explain geometrical illusions produced by crossing lines. Perception and Psychophysics 3, 174–176.

Coffin, S. 1978. Spatial frequency analysis of block letters does not predict experimental confusions. Perception and Psychophysics 23, 69–74.

Coren, S. 1969. Brightness contrast as a function of figure-ground relations. Journal of Experimental Psychology 80, 517–524.

Coren, S. 1970. Lateral inhibition and geometric illusions. Quarterly Journal of Experimental Psychology 22, 274–278.

Coren, S. and Girgus, J.S. 1978. Seeing is deceiving: The psychology of visual illusions. Hillsdale, New Jersey: Lawrence Erlbaum Associates.

Davies, P. 1973. The role of central processes in the perception of visual afterimage fragmentation. British Journal of Psychology 64, 325–338.

Dember, W.N. and Purcell, D.G. 1967. Recovery of masked visual targets by inhibition of the masking stimulus. Science 157, 1335–1336.

DeValois, R.L. and Pease, P.L. 1971. Contours and Contrast: Response of monkey lateral geniculate nucleus cells to luminance and color figures. Science 171, 694–696.

Eriksson, E.S. 1970. A field theory of visual illusions. British Journal of Psychology 61, 451–466.

Evans, C.R. 1965. Some studies of pattern perception using a stabalized retinal image. British Journal of Psychology 56, 121-133.

Fehrer, E., and Raab, D. 1962. Reaction time to stimuli masked by metacontrast. Journal of Experimental Psychology 63, 143-147.

Festinger, L., Coren, S., and Rivers, G. 1970. The effect of attention on brightness contrast and assimilation. American Journal of Psychology 83, 189-207.

Fox, R. and Lehmkuhle, S. Contour interaction in visual space: Depth separation and visual masking. (Report #N14-1101 78C-0001). Vanderbilt University, Nashville, Tennessee: Department of Psychology, July 1978.

Ganz, L. 1966a. Mechanism of the figural aftereffects. Psychology Review 73, 128-150.

Ganz, L. 1966b. Is the figural aftereffect an aftereffect? A review of its intensity, onset, decay and transfer characteristics. Psychology Bulletin 66, 151-165.

Gilchrist, A. 1977. Perceived brightness depends on perceived spatial arrangement. Science 195, 185-187.

Gogel, W.C., and Mershon, D.H. 1969. Depth adjacency in simultaneous contrast. Perception and Psychophysics 5, 13-17.

Gogel, W.C., and Newton, R.F. 1975. Depth adjacency and the rod-frame illusion. Perception and Psychophysics 18, 163-171.

Goldstein, M.B., and Weintraub, D.J. 1973. The parallel-less Paggendorff: Virtual contours put the illusion down but not out. Perception and Psychophysics 11, 353-354.

Graham, N. Spatial frequency channels in human vision: Detecting edges without edge detectors. In C.S. Harris, (Ed.), Visual coding and adaptability. Hillsdale, New Jersey: Lawrence Erlbaum Associates, in press.

Greenwood, R.E. 1973. Visibility of structured and unstructured images. Journal of the Optical Society of America 63,226-231.

Georgeson, M.A. 1976. Psychophysical hallucinations of orientation and spatial frequency. Perception 5, 99-111.

Growney, R. 1978. Metacontrast as a function of the spatial frequency composition of the target and mask. Vision Research 18, 1117-1124.

Heggelund, P., and Hohmann, A. 1976. Long term retention of the "Gilinsky effect." Vision Research 16, 1015-1017.

Henning, G.B., Hertz, B.G. and Broadbent, D.E. 1975. Some experiments bearing on the hypothesis that the visual system analyzes spatial patterns in independent bands of spatial frequency. Vision Research 15, 887-897.

Hernandez, L.L., and Letton, L.A. 1977. Metacontrast as measured under a signal detection method. Perception 6, 695-702.

Hogden, J.H., and DiLollo, V. Practice induced decrement of suppression in metacontrast and apparent motion. In press.

Jenkins, B., and Ross, J. 1977. McCullough effect depends on perceptual organization. Perception 6, 399-400.

John, E.R., and Schwarz, E.L. 1978. The neurophysiology of information processing and cognition. Annual Review of Psychology 29, 1-29.

Jones, P.D., and Holding, D.H. 1975. Extremely long-term persistence of the McCullough effect. Journal of Experimental Psychology: Human Perception and Performance 1, 323-327.

Julesz, B. 1971. Foundations of cyclopean perception. Chicago: University of Chicago Press.

Kaniza, G. 1975. Some new demonstrations of the role of structural factors in brightness contrast. In S.Ertel, L. Kemmler, and M.Stadler (Eds.), Gestaltheorie in der Moderne Psychologie. Stuttgart: Steinkopf.

Kaniza, G. 1976. Subjective contours. Scientific American, 234, 48-52.

Kohler, W., and Wallach, H. 1944. Figural aftereffects: An investigation of visual processes. Proceedings of the American Philosophical Society, 88, 269-357.

Kolers, P.A. 1964. The illusion of movement. Scientific American, 211, 98-106.

Kolers, P.A. 1970. The role of shape and geometry in picture recognition. In B.S. Lipkin, and A. Rosenfeld, (Eds.), Picture processing and psychopictories. New York: Academic Press.

Kornorski, J. 1967. Integrative active of the brain. Chicago: University of Chicago Press.

Koppitz, W.J. 1957. Mach bands and retinal interaction. Unpublished doctoral dissertation, Ohio State University.

Land, E.H. Nov. 1977. The retinex theory of color vision. Scientific American, 108-128.

Leeper, R. 1935. A study of a neglected portion of the field of learning--the development of sensory organization. Journal of Genetic Psychology, 46, 41-75.

MacKay, D.R. and Mackay V. 1975. What causes decay of pattern-contingent chromatic aftereffects. *Vision Research,* 15, 462-464.

Mayzner, M.S., and Tresselt, M.E. 1970. Visual information processing and sequential inputs: A general model for sequential blanking displacement, and overprinting phenomena. *Annals of New York Academy of Science,* 169, 599-618.

Mershon, D.H. 1972. Relative contribution of depth and directional adjacency to simultaneous whiteness contrast. *Vision Research,* 12, 969-979.

Mershon, D.H., and Gogel, W.C. 1970. The effect of stereoscopic cues on perceived whiteness. *American Journal of Psychology,* 85, 55-67.

Mize, R.R., and Murphy, E.H. 1973. Selective visual experience fails to modify receptive field properties of rabbit striate cortex neurons. *Science,* 180, 320-323.

Motokawa, K. 1950. Field of retinal induction and optical illusion. *Journal of Neurophysiology,* 13, 413-26.

Motokawa, K., and Akita, M. 1957. Electrophysiological studies of the field of retinal induction. *Psychologia,* 1, 10-16.

Nachmias, J. and Weber, A. 1975. Discrimination of simple and complex gratings. *Vision Research,* 15, 217-222.

Obanai, T. 1977. *Perception, learning, and thinking--Psychophysiological induction theory.* Tokyo: Hokuseido Press.

Poggio, T. 1979. Trigger features or Fourier analysis in early vision: A new point of view. Paper presented at the University of Texas conference on feature detection. Austin, Texas.

Pollack, I. 1972. Visual discrimination of "unseen" objects: forced choice testing of Mayzner-Tresselt sequential blanking effects. *Perception and Psychophysics,* 11, 121-128.

Pomerantz, J.R. 1978. Are complex visual features derived from simple areas? In E. L. J. Leeuwenber, and H. F. J. M. Buffart, (Eds.), *Formal theories of visual perception,* New York: Wiley.

Pomerantz, J.R., Sager, L.C., and Stoever, R. J. Perception of wholes and their component parts: Some configurational superiority effects. *Journal of Experimental Psychology: Human Perception and Performance,,* 3, 422-435.

Prinzmetal, W., and Banks, W.P. 1977. Good continuation affects visual detection. Personal Communications.

Pritchard, R.M., Heron, W., and Hebb, D.O. 1960. Visual perception approached by the method of stabalized images. *Canadian Journal of Psychology,* 14, 67-77.

Ratliff, F. 1965. Mach bands: Quantitative studies on neural networks in the retina. San Francisco: Holden-Day.

Riggs, L.A., White, D.D., and Eimas, P.D. 1974. Establishment and decay of orientation-contingent aftereffects of color. Perception and Psychophysics, 16, 535-542.

Robinson, D.N. 1966. Distribution of visually masked stimuli. Science, 154, 157-158.

Robinson, G.M., and Moulton, J. 1978. A delayed induced-motion illusion. Perception, 7, 85-89.

Robinson, J.O. 1972. The psychology of visual illusion. London: Hutchinson University Library.

Schiller, P.H., and Smith, M.C. 1966. Detection in metacontrast. Journal of Experimental Psychology, 71, 32-39.

Sharpe, L.T., and Teas, R.C. 1978. Contour specificity of the McCollough effect. Perception and Psychophysics, 23, 451-458.

Sherrington, C.S. 1906. Integrative Activity of the nervous system. New Haven: Yale University Press.

Skowbo, D., Gentry, T., Timney, B., and Morant, R.B. 1974. The McCollough effect: influence of several kinds of visual stimulation on decay rate. Perception and Psychophysics, 16, 47-49.

Smith, A.T., and Over, R. 1977. Orientation masking and the tilt illusion with subjective contours. Perception, 6, 441-447.

Stetcher, S., Sigel, C., and Lange, R.V. 1973. Composite adaptation and spatial frequency interaction. Vision Research, 13, 2527-2531.

Stromeyer, C.F.III, and Klein, S. 1975. Evidence against narrow-band spatial frequency channels in human vision: The detectability of frequency modulated gratings. Vision Research, 15, 899-910.

Stromeyer, C.F.III, and Mansfield, R.J.W. 1970. Colored aftereffects produced with moving edges. Perception and Psychophysics, 7, 108-114.

Stryker, M.P., and Sherk, H. 1975. Modification of cortical orientation selectivity in the cat by restricted visual experience. A reexamination. Science, 190, 904-906.

Thompson, P.G., and Movshon, J.A. 1978. Storage of spatially specific threshold elevation. Perception, 7, 65-73.

Timney, B.N., and MacDonald, C. 1978. Are curves detected by "curvature detectors?" Perception, 7, 51-64.

Tolhurst, D.J. 1972. Adaptation to square wave gratings: Inhibition between spatial frequency channels in human visual system. Journal of Psychiology: London, 226, 231-248.

Tolhurst, D.J., and Barfield, L.P. 1978. Interaction between spatial frequency channels. Vision Research, 18, 951-958.

Towe, A.L. 1975. Notes on the hypothesis of columnar organization in somatosensory cerebral cortex. Brain, Behavior, and Evolution, 11, 16-47.

Uttal, W.R. 1970. On the psysiological basis of masking with dotted visual noise. Perception and Psychophysics, 7, 321-327.

Uttal, W.R. 1973. The psychobiology of sensory coding. New York: Harper and Row.

Uttal, W.R. in press. A taxonomy of visual processes. Hillsdale, New Jersey: Lawrence Erlbaum Associates.

Walker. D. 1978. Binocular Rivalry: Central or peripheral selective processes? Psychology Bulletin, 85, 376-389.

Weisstein, N., and Harris, C.S. 1974. Visual detection of line segments: An objectsuperiority effect. Science, 186, 752-755.

White, K.D. 1978. Studies of form-contingent color aftereffects. In J.C. Armington, J. Krauskopf, and B.R. Wooten, (Eds.), Visual psychophysics and physiology, New York: Academic Press.

Zomansky, H.S., and Corwin. 1976. Word length and visual noise texture in backward masking. Perception, 5, 211-215.

Bio-mathematics

Managing Editor: S. A. Levin

Springer-Verlag
Berlin
Heidelberg
New York

Volume 8
A. T. Winfree

The Geometry of Biological Time

1979. 290 figures. XIV, 530 pages
ISBN 3-540-09373-7

The widespread appearance of periodic patterns in nature reveals that many living organisms are communities of biological clocks. This landmark text investigates, and explains in mathematical terms, periodic processes in living systems and in their non-living analogues. Its lively presentation (including many drawings), timely perspective and unique bibliography will make it rewarding reading for students and researchers in many disciplines.

Volume 9
W. J. Ewens

Mathematical Population Genetics

1979. 4 figures, 17 tables. XII, 325 pages
ISBN 3-540-09577-2

This graduate level monograph considers the mathematical theory of population genetics, emphasizing aspects relevant to evolutionary studies. It contains a definitive and comprehensive discussion of relevant areas with references to the essential literature. The sound presentation and excellent exposition make this book a standard for population geneticists interested in the mathematical foundations of their subject as well as for mathematicians involved with genetic evolutionary processes.

Volume 10
A. Okubo

Diffusion and Ecological Problems: Mathematical Models

1980. 114 figures, 6 tables. XIII, 254 pages
ISBN 3-540-09620-5

This is the first comprehensive book on mathematical models of diffusion in an ecological context. Directed towards applied mathematicians, physicists and biologists, it gives a sound, biologically oriented treatment of the mathematics and physics of diffusion.

Journal of

Mathematical

Biology

ISSN 0303-6812   Title No. 285

Editorial Board:
H.T.Banks, Providence, RI; **H.J.Bremermann,**
Berkeley, CA; **J.D.Cowan,** Chicago, IL; **J.Gani,**
Canberra City; **K.P.Hadeler** (Managing Editor),
Tübingen; **S.A.Levin** (Managing Editor), Ithaca, NY;
D.Ludwig, Vancouver; **L.A.Segel,** Rehovot; **D.Varjú,**
Tübingen

Advisory Board: M. A. Arbib, W.Bühler, B.D.Coleman,
K.Dietz, F.A.Dodge, P.C.Fife, W.Fleming, D.Glaser,
N.S.Goel, S.P.Hastings, W.Jäger, K.Jänich, S.Karlin,
S.Kauffman, D.G.Kendall, N.Keyfitz, B.Khodorov,
J.F.C.Kingman, E.R.Lewis, H.Mel, H.Mohr,
E.W.Montroll, J.D.Murray, T.Nagylaki, G.M.Odell,
G.Oster, L.A.Peletier, A.S.Perelson, T.Poggio,
K.H.Pribram, J.M.Rinzel, S.I.Rubinow, W.v.Seelen,
W.Seyffert, R.B.Stein, R.Thom, J.J.Tyson

Springer-Verlag
Berlin
Heidelberg
New York

The **Journal of Mathematical Biology** publishes papers
in which mathematics leads to a better understanding
of biological phenomena, mathematical papers inspired
by biological research and papers which yield new experimental data bearing on mathematical models. The
scope is broad, both mathematically and biologically
and extends to relevant interfaces with medicine,
chemistry, physics and sociology. The editors aim to
reach an audience of both mathematicians and
biologists.

Subscription information and sample copy
upon request.

Lecture Notes in Biomathematics